Genetic Studies of Fish: I

Papers by
James E. Wright, George J. Ridgway,
William J. Morrison et al.

MSS Information Corporation
655 Madison Avenue, New York, N.Y. 10021

Library of Congress Cataloging in Publication Data
Main entry under title:

Genetic studies of fish.

   1. Fish genetics--Addresses, essays, lectures.
I. Wourms, John P. [DNLM: 1. Fishes--Collected works.
2. Genetics--Collected works. QL615 G328 1974]
QL639.1.G46   597'.01'5   74-516
ISBN 0-8422-7207-0 **(v. 1)**
     0-8422-7177-5 (V.2)

TABLE OF CONTENTS

34529

CREDITS AND ACKNOWLEDGEMENTS

Avise, John C.; and G. Barrie Kitto, "Phosphoglucose Isomerase Gene Duplication in the Bony Fishes: An Evolutionary History," *Biochemical Genetics*, 1973, 8:113-132.

Davisson, Muriel Trask; J.E. Wright; and Louisa M. Atherton, "Centric Fusion and Trisomy for the LDH-B Locus in Brook Trout, *Salvelinus fontinalis,*" *Science*, 1972, 178:992.

Eckroat, Larry R., "Allele Frequency Analysis of Five Soluble Protein Loci in Brook Trout, *Salvelinus fontinalis* (Mitchill)," *Transactions of the American Fisheries Society*, 1973, No. 2:335-340.

Eckroat, Larry R., "Lens Protein Polymorphisms in Hatchery and Natural Populations of Brook Trout, *Salvelinus fontinalis* (Mitchill)," *Transactions of the American Fisheries Society*, 1971, No. 3:527-536.

Hershberger, William K., "Some Physicochemical Properties of Transferrins in Brook Trout," *Transactions of the American Fisheries Society*, 1970, 99:207-218.

Morrison, William J., "Nonrandom Segregation of Two Lactate Dehydrogenase Subunit Loci in Trout," *Transactions of the American Fisheries Society*, 1970, 99:193-206.

Ridgway, George J.; Stuart W. Sherburne; and Robert D. Lewis, "Polymorphism in the Esterases of Atlantic Herring," *Transactions of the American Fisheries Society*, 1970, 99:147-151.

Whitt, Gregory S.; William F. Childers; John Tranquilli; and Michael Champion, "Extensive Heterozygosity at Three Enzyme Loci in Hybrid Sunfish Populations," *Biochemical Genetics*, 1973, 8:55-72.

Wright, James E. Jr.; and Louisa M. Atherton, "Polymorphisms for LDH and Transferrin Loci in Brook Trout Populations," *Transactions of the American Fisheries Society*, 1970, 99:179-192.

PREFACE

There has been a recent surge of interest in fish genetics by those concerned with fish farming or the exploitation of fish populations in the wild. In large measure, this interest has been directed toward practical matters, such as the development of a superior breed of fish. Most agree that the desired fish would meet the following criteria: resistance to disease, good food conversion ratio, palatability, quick growth, and dependable offspring. The poultry and livestock industries have worked for scores of years to achieve similar standards, often without knowledge of genetic principles. However, indications are that fish may be genetically manipulated in a way not yet seen on a wide scale anywhere else.

The key to this possible manipulation lies in the fact that certain fish species undergo a bizarre type of fertilization and early embryological development. The process, termed gynogenesis, is a form of pathenogenesis whereby an ovum is activated by sperm, but without the contribution of the paternal genome to the offspring. Amphibia are the only other natural population of gynogenetic vertebrates that have been identified.

Although the complex cytological mechanisms of gynogenesis are just beginning to be understood, the end result is clear: an offspring which is a genetic duplicate of its mother. This phenomena may be utilized to improve strains of fish. Genetic manipulation appears feasible, since artificial gynogenesis can be carried out by reacting eggs with irradiated sperm. These sperm, although they can activate the egg, have had their genome destroyed by the irradiation. Thus they contribute nothing genetically to the offspring, and genetic characters selected in the female can be maintained. Also important to such manipulation is the low incidence of crossing over in fish.

This two-volume collection of readings covers the period since 1969, and emphasizes key topics. The present volume, *Genetic Studies of Fish: I*, includes studies on the biochemical genetics of fish.

A companion volume, *Genetic Studies of Fish: II*, contains research on gynogenesis and other aspects of genetic investigation important to effective breeding of fish. Cytogenetics and more general types of genetic approaches are also covered.

Ronald T. Acton, Ph.D.

January, 1974

Polymorphisms for LDH and Transferrin Loci in Brook Trout Populations[1]

JAMES E. WRIGHT, JR. AND LOUISA M. ATHERTON

ABSTRACT

Seven hatchery populations and seven of eight wild populations of brook trout sampled were polymorphic for both the LDH-B and the transferrin loci. The degree of variations of allele frequencies as well as of the amount of heterozygosity was generally greater among hatchery fish than those from natural populations. Calculations of interpopulation heterogeneities of allele frequency differences permitted distinguishing all of the hatchery strains, but not all of the small wild populations, from each other. While gross intra-population heterogeneity was found, an assessment of the breeding structure of wild brook trout in a stream was not possible.

Electrophoretic variants were found in wild populations for two other LDH polypeptides and for one other serum protein. From assumptions that five loci were polymorphic in 13 total loci studied, the estimate of proportion of polymorphic loci in wild brook trout is 38.4%. The similar estimate for hatchery strains is 15.4%.

INTRODUCTION

Allele frequency analyses of protein polymorphisms in fish populations can serve at least three interests. One of these is the assessment of the discreteness of stocks or races of fish species, of particular importance with marine fisheries (Marr and Sprague, 1963; Parrish, 1964). Of broader biological interest are the measurements of amounts of genetic variation in populations such as has been done in *Drosophila* (Hubby and Lewontin, 1966; Lewontin and Hubby, 1966), mice (Selander and Yang, 1969), catostomid fish (Koehn and Rasmussen, 1967), and man (Harris, 1966; Harris *et al.*, 1968) and the determination of breeding structure in natural populations as done in mice (Petras, 1967; Selander and Yang, 1969).

We have attempted to contribute to these interests by surveying allele frequencies at two loci for which definite genetic bases have been determined in the brook trout, *Salvelinus fontinalis*. One of these loci is that controlling the serum protein transferrin; the other is that controlling one of the five subunits,

[1] This work was supported in part by NSF Grant GB4624.

LDH-B, involved in formation of one of the three series of the isoenzymatic protein lactate dehydrogenase (Morrison and Wright, 1966). Allele frequencies can be determined directly for both loci since all possible genotypes are distinguishable by zone electrophoresis. The discovery of new variants at these and other loci for use in our genetic analyses formed an additional interest in this study. A variant at the LDH-A locus was sought particularly in order to obtain brook trout doubly heterozygous for the A and B loci. Such a variant would permit further investigation of the pseudolinkage phenomenon found when utilizing the LDH-A from lake trout as a variant to obtain doubly heterozygous splake hybrids (Morrison, 1970).

Transferrin polymorphisms have been described in a number of fish species (Fine, *et al.*, 1965), in scombroid fishes (Barrett and Tsuyuki, 1967; Fujino and Kang, 1968), and in plaice, *Pleuronectes platissa*, (deLigny, 1967). Gene frequency analyses of transferrin polymorphisms in gadoid fishes were used to distinguish races of these fishes (Møller and Naevdal, 1966; Jamieson and Jones, 1967). Polymorphisms for certain LDH loci have been described in Atlantic populations of whiting, *Merluccius bilinearis* (Markert and Faulhaber, 1965), and of herring, *Clupea harengus* (Odense *et al.*, 1966). Goldberg (1966) and Morrison and Wright (1966) have reported polymorphisms for the LDH-B locus in hatchery brook trout populations.

In the present study two types of brook trout populations were sampled—those from a number of hatcheries in the Northeast and those primarily from small, isolated streams in central Pennsylvania in which the brook trout is a native species. The first group represents relatively artificial allopatric populations of large effective breeding units, since hundreds of thousands of fish are produced each year by randomly mating hundreds of brood fish maintained at each hatchery. The second group represents natural allopatric populations

10

of relatively small effective breeding units of unknown structure since the streams have small pools separated by long shallow riffles or by rock ledge barriers.

Transferrin phenotypes were determined by acrylamide gel disc electrophoresis on a Canalco Model 12 apparatus using a gel concentration of 5% (Davis, 1964). Blood was taken from yearling or mature fish by cardiac puncture, allowed to clot, centrifuged to remove cell debris, the serum decanted and stored at –20 C. Tails of fingerlings or other small fish were clipped and blood collected from the caudal artery into hematocrit tubes. The tubes were centrifuged, cut off at the interface of serum and clot, and the serum section sealed at both ends for frozen storage.

LDH phenotypes were determined by horizontal starch gel electrophoresis using vitreous humor or eye homogenates applied to filter paper wicks. The procedures have been described by Morrison (1970). In the earlier stages of the work when acrylamide gel disc electrophoresis was employed (Morrison and Wright, 1966), all phenotypes were not distinguishable. However, remnant samples of all but one population were available to rerun on starch gels. Vitreous humor was collected from larger fish by means of a hypodermic syringe fitted with a one-inch, 18 gauge needle. Whole eyes of fingerlings or fry were dissected and homogenized in two volumes of glass distilled water in a glass Teflon homogenizer.

Vitreous humor and serum samples were collected from yearling or mature fish of the hatchery strains at the hatchery site. The fish were anesthesized with 3-methyl-1-pentyn-3-ol and were not killed. Fingerlings and fry populations were transported to the Benner Spring Fish Research Station and held until sampled. The strains, sources, and those responsible for aid in the collection, respectively, of the hatchery strains were as follows:

a. The Pennsylvania strain, a derivative of

11

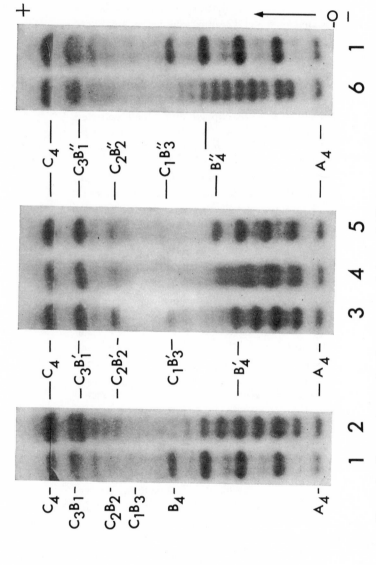

FIGURE 1.—Comparative starch gel zymograms of the six LDH phenotypes involving the B, B', and B" alleles—in vitreous humor of brook trout. (1 = BB; 2 = BB; 3 = B'B'; 4 = B'B"; 5 = B"B"; 6 = BB"; 0 = origin).

the Paradise and Trexler hatchery strains and maintained as a closed, randomly bred strain for many generations; Reynoldsdale, Pennsylvania; Mr. George Magargel.

b. The Gilbert strain; Warren, New Hampshire; Mr. C. B. Corson and Mr. Charles Evans.

c. The Hackettstown (New Jersey)—Paradise strain, introduced into West Virginia hatcheries and maintained as separate, closed populations for three generations before our collection; Ridge, and Edray, West Virginia; Mr. Harvey Beall and Mr. David Cochran.

d. The Coleville strain, a semi-wild strain introduced from an eastern hatchery numbers of years ago into a lake on the property of the Coleville Indian Agency, Nespelem, Washington, the eggs of which have been obtained by the U. S. Fish and Wildlife Service; National Fish Hatchery, White Sulphur Springs, West Virginia; Mr. Chester Ambrose.

e. The Port Arthur strain (limited sample); Normandale, Ontario; Dr. Hugh McCrimmon.

f. The S V strain, a second generation synthetic strain from mass breeding of survivors of furunculosis disease screened from various eastern hatchery strains; Rome, New York; Mr. Neil Ehlinger.

Wild brook trout were collected by sport fishing or electrofishing and generally only fish three inches long or longer were sampled since we were interested in saving any unusual types for breeders. The fish were marked with numbered metal jaw tags as the vitreous humor was collected, then transported and held at the Benner Spring Station. They were first typed for LDH and then the surviving fish were bled for serum samples. The sources of the fish were:

a. The headwaters of Marsh Creek in Centre County, Pennsylvania. Samples were taken from the uppermost part of the stream (Section I) during the spring and fall of 1967 and the spring of 1968. The downstream section (II), separated by over 200 yards from Section I, was sampled during the summer and fall of 1968 after further division into subsections. Each subsection was 20–30 yards in length and arbitrarily divided on the basis of natural barriers in the stream.

b. Five small streams which drain into Catawissa Creek, Columbia County, Pennsylvania. Catawissa Creek has been polluted by acid coal mine drainage for a century and no fish live in this stream. Thus what probably was one interbreeding population originally, has been subdivided into the subpopulations of each tributary stream.

c. Limited numbers of samples were obtained from Maiden Canyon Creek, Montana, from a population introduced from the East numbers of years ago, and from Squabble Brook near Caanan, Connecticut, while pursuing avocational interests.

RESULTS AND DISCUSSION

LDH and Transferrin Phenotypes and Their Genetic Bases

The nomenclature utilized by Morrison and Wright (1966) to identify five loci specifying five subunits involved in the production of three series of LDH isozymes in brook trout has been followed in this report. The LDH-A and -B loci specify subunits A and B which produce isozymes occurring in practically all tissues; we have operationally designated this as the ubiquitous series. The LDH-C locus specifies the retinal or C subunit which combines with A and with B subunits to produce isozymes occurring in the eye and brain, designated the eye series. The D and E subunits specified by the LDH-D and -E loci produce isozymes restricted to muscle tissues, and hence designated the muscle series.

The six phenotypes involving all possible genotypic combinations of three alleles found at the B locus (B, B′, and B″) are distinguishable on starch gel zymograms of vitreous humor samples as shown in Figure 1. This distinction is apparent not only for the homo-

14

TABLE 1.—*Genetic analysis of LDH-B variants*

Family	Strain	Parental phenotypes Female	Male	BB	BB'	BB"	B'B'	B'B"	B"B"	P of chi-square
M-289	12	BB"	BB"	12 (12.5)		26 (25)			12 (12.5)	.95
O-303	0	B'B"	B'B"				31 (27.75)	53 (55.5)	27 (27.75)	>.70
O-49	12	BB	B'B"		50 (52)	54 (52)				.70
M-332	0	B'B"	BB"		51 (42.5)	44 (42.5)		44 (42.5)	31 (42.5)	.20

15

and heterotetramers involving the A and B subunits in the ubiquitous series, but also for those heterotetramers of the eye series involving the various B and the single C subunits. Thus use of vitreous humor permits typing fish for polymorphisms of the LDH-B, -A and -C loci without killing or disfiguring them. Its use has the disadvantage that one cannot score for variants at the LDH-D and -E loci.

Vitreous humor of the same six LDH phenotypes shown as starch gel zymograms in Figure 1 were subjected to acrylamide gel disc electrophoresis and the acrylamide gel zymograms are compared in Figure 2. It is apparent that the two homozygotes, B′B′ and B″B″, and the heterozygote B′ B″ are indistinguishable from each other; neither are the B B′ and B B″ heterozygotes distinguishable under the conditions of electrophoresis employed. It should be pointed out also that the muscle series of LDH isozymes are not resolved with acrylamide gel electrophoresis. Therefore, LDH types which were scored on acrylamide gels in the earlier phases of our population surveys had to be reanalyzed utilizing starch gels.

Genetic analysis was reported earlier (Morrison and Wright, 1966) which showed that the B and B′ are codominant alleles at an autosomal locus. Data are presented in Table 1 which compare the observed and expected frequencies of offspring from parents of known genotypes in different inbred line families maintained at the Benner Spring Station. The results show clearly that the B″ variant is allelic to B and B′. In addition, true-breeding strains of B″ B″, as well as of B′B′ and B B, homozygotes have been established. Preliminary data have been reported (Wright and Atherton, 1968) which were interpreted as evidence that the B′ and B″ alleles are pseudoallelic. This is, when B′B″ males were mated to B B females, B B progeny appeared in frequencies which were influenced by a segregating genetic element in the male

16

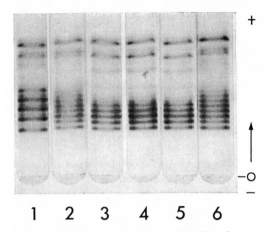

FIGURE 2.—Comparative disc acrylamide gel zymograms of the six LDH phenotypes showing inability to distinguish homozygotes or heterozygotes involving the B' and B" alleles. (Numbers refer to the same phenotypes listed in Figure 1.)

parents. This evidence, plus the high frequency of the B allele in most populations reported here leads to the interpretation that the B is the wild type allele.

The genetic basis for polymorphism at the one transferrin locus (Tf) in brook trout was determined in this laboratory by Hoffman (1966) to be three codominant autosomal alleles. The three alleles condition three electrophoretically distinct transferrins, designated A, B and C in decreasing order of anodal migration, and the six possible phenotypes (also genotypes) are distinguishable on acrylamide gel electrophorograms. Hershberger (1970) has illustrated and has described aspects of these six transferrin phenotypes. The C allele is considered the wild type on the basis of its frequency in populations studied.

Evidence for the independent inheritance of the LDH-B and Tf loci is shown in Table 2. The progenies of doubly heterozygous males are shown to appear in frequencies closely approximating those expected on the basis of independent assortment. Although not shown

17

here, very small numbers in progenies from several matings in which females were doubly heterozygous provide collective evidence that there is no sexual difference in the independent transmission of alleles at the two loci. The fact that the two loci are genetically independent is another advantage for their use in allele frequency analyses of populations.

Phenotype and Allele Frequencies in Hatchery Populations

The extensive polymorphism of the LDH-B locus in brook trout from all hatcheries surveyed is evident from the data presented in Table 3. Allele frequencies were calculated from the array of phenotypes observed for each year class sampled at each hatchery. These frequencies were then utilized to calculate expected values for each phenotype and it may be seen that every population exhibits Hardy-Weinberg equilibrium. However, in two populations, Warren, New Hampshire, and Ridge, West Virginia, where the previous generation of breeders had been sampled, use of the allele frequencies of the parental group to calculate expected phenotypes in the offspring population indicates lack of stability from one generation to the next. Although there may be a question as to whether the allele frequencies of the parents were adequately sampled, it might be suggested that excess of heterozygotes and deficiency of homozygotes in the offspring generation cause the poor fit between observed and expected when considered between generations. This may indicate heterozygote superiority and could be the mechanism for maintenance of the high degree of polymorphism although the evidence is subtle. Unfortunately, the B″ allele was not distinguishable in the Reynoldsdale, Pennsylvania breeders in 1966 because the phenotypes were determined by acrylamide gel electrophoresis. It appears, however, that there is little change in allele frequency from generation to generation in this hatchery if one compares the B allele frequency with the

18

total of the two variant alleles. Further, the low, stable frequencies of the variant alleles argue against heterozygote superiority in this hatchery.

It is noteworthy that the B″ allele occurs in the four hatchery strains which have the Paradise strain as at least a part of their heritage. The newly derived synthetic strain from Rome, New York, possesses this allele in very high frequency. Indeed, the frequencies of the variant alleles total 78% in this strain, the only one in which the frequency of the wild type B alleles is less than 50%.

All populations of the hatchery strains of brook trout exhibit a high degree of polymorphism for the transferrin locus as shown by the array of phenotype and allele frequencies presented in Table 4. When the allele frequencies calculated from the observed phenotypes were used to obtain the expected values, every population was shown to be in Hardy-Weinberg equilibrium. In the Warren, New Hampshire, and particularly the Reynoldsdale, Pennsylvania, hatcheries the allele frequencies remained quite stable over generations. This is shown by the acceptable fits between observed phenotype frequencies in offspring generations and those expected on the basis of random combinations of parental gametes. Since brood fish are often held over for another breeding season at the Reynoldsdale hatchery, we have used the allele frequencies of the 1966 breeders to calculate the expected frequencies of the 1968 fingerlings spawned in 1967. Although the Chi-square value was too high to be acceptable as chance occurrence, the two AA types contribute inordinately to the Chi-square value since the expected frequency was less than 0.5. Without this value there would have been an acceptable fit.

The two AA phenotypes were unexpected also on the basis that Hoffman (1966) showed that this homozygous type was associated with lethality. Their occurrence here could be evidence for another, non-lethal A isoallele pro-

TABLE 3.—*LDH-B phenotype and allele frequencies in hatchery brook trout populations*

Population		Total	Phenotypes						P of chi-square	Alleles		
			BB	BB'	BB"	B'B'	B'B"	B"B"		B	B'	B"
Pennsylvania:												
Reynoldsdale breeders, 1966*	Observed	127	107	19	—	1	—	—		.92	.08	—
	Expected		107	19	—	1	—	—	.50			
Reynoldsdale yearlings, 1968	Observed	209	181	9	18	—	—	1		.93	.02	.05
	Expected		181	8	19	—	—	1	>.70			
Reynoldsdale fingerlings, 1968	Observed	331	280	25	25	—	—	1		.92	.04	.04
	Expected		281	24	24	0.5	0.5	1	>.70			
New Hampshire:												
Warren (Gilbert) yearlings, 1966	Observed	49	27	18	—	4	—	—		.73	.27	—
	Expected		26	19	—	4	—	—	.70			
Warren (Gilbert) fingerlings, 1968	Observed	151	58	74	—	19	—	—		.63	.37	—
	Expected		60	70	—	21	—	—	.50			
	Expected[a]		80	60	—	11	—	—	<.01			
West Virginia:												
Ridge breeders, 1967	Observed	103	44	39	2	15	2	1		.63	.34	.03
	Expected		40	44	4	13	2	1	.50			
Ridge fingerlings, 1968	Observed	249	84	112	20	21	11	1		.60	.33	.07
	Expected		90	99	21	27	11	1	>.30			
	Expected[b]		98	107	9	30	5	1	<.001			
Edray 2-yr-olds, 1968	Observed	166	38	84	7	30	7	—		.50	.46	.04
	Expected		41	77	7	35	6	—	>.50			
U. S. Fish and Wildlife Service:												
Coleville Indian 2-yr-olds, 1968	Observed	50	38	11	—	1	—	—		.87	.13	—
	Expected		38	12	—	1	—	—	>.40			
Ontario:												
Normandale yearlings, 1966	Observed	13	6	5	—	2	—	—		.65	.35	—
	Epected		6	6	—	1	—	—	>.50			
New York:												
Rome fry, 1969	Observed	263	13	57	31	63	81	18		.22	.50	.28
	Expected		12	57	32	66	74	21	>.70			

* Phenotypes involving B" allele were indistinguishable from those involving B' by polyacrylamide disc electrophoresis used with this one population.
a Expected values calculated on basis of allele frequencies in 1966 yearlings.
b Expected values calculated on basis of allele frequencies in 1967 breeders.

ducing a transferrin with identical electrophoretic mobility. This hypothesis is under
investigation since several AA homozygotes
have been found in one inbred line at the
Benner Spring Station. If the A allele is a
non-lethal isoallele, this could account for the
relatively high, stable frequency in which it
occurs uniquely in the Pennsylvania strain
and, in a remarkably similar frequency, in the
semi-wild Coleville Indian strain. If it is not,
perhaps Hershberger's (1970) evidence that
heterozygotes are superior in iron binding
and release could account for a balanced
polymorphism. However, there is no evidence
in the populations studied that the heterozygotes involving the A allele are in excess.
Indeed, there is no consistent evidence for
superiority of heterozygotes for any Tf alleles
in any of these hatchery populations.

In the Ridge, West Virginia, hatchery the
allele frequencies of parental and offspring
generations are similar. However, in the
fingerling offspring there is a poor fit between observed phenotypes and those expected
on the basis of parental allele frequencies. The
excess of homozygotes over those expected
might suggest a degree of inbreeding in this
hatchery, if the small sample of breeders
adequately reflected the true allelic frequencies
in breeders used in 1967.

Intra- and Inter-Hatchery Heterogeneity

The significance of differences among year
classes within hatchery strains as well as of
differences between hatchery strains can be
shown by utilizing actual allele frequencies
to calculate heterogeneity Chi-square values
(contingency test). The intra-hatchery tests
for heterogeneity for each of the three
hatcheries in which different year classes were
sampled show that the differences among
year classes were not significant for either
locus (Table 5). From this test one may
conclude that there is a stability of allele
frequencies from generation to generation, a
conclusion somewhat at variance with that

21

TABLE 4.—*Transferrin phenotype and allele frequencies in hatchery brook trout populations*

Population			Phenotypes						P of chi-square	Alleles		
		Total	AA	AB	AC	BB	BC	CC		A	B	C
Pennsylvania:												
Bellefonte 3-year-olds, 1963	Observed	77	—	5	4	4	25	39		.06	.25	.69
	Expected		—	2	6	5	27	37	>.10			
Reynoldsdale breeders, 1966	Observed	130	—	4	7	14	43	62		.04	.29	.67
	Expected		—	3	8	11	50	58	.40			
Reynoldsdale yearlings, 1968	Observed	101	—	5	6	11	43	36		.05	.35	.60
	Expected		—	4	7	12	42	36	>.80			
	Expected[a]		2	2	5	9	39	45	>.10			
Reynoldsdale fingerlings, 1968	Observed	294	2	7	19	25	129	112		.05	.32	.63
	Expected		—	9	19	30	119	117	>.20			
	Expected[a]		—	10	16	25	114	132	.01			
New Hampshire:												
Warren (Gilbert) yearlings, 1966	Observed	49	—	—	—	7	22	20		—	.37	.63
	Expected		—	—	—	6	23	20	>.80			
Warren (Gilbert) fingerlings, 1968	Observed	145	—	—	—	12	73	60		—	.33	.67
	Expected		—	—	—	16	65	64	>.10			
	Expected[b]		—	—	—	20	69	58	.05			
West Virginia:												
Ridge breeders, 1967	Observed	29	—	—	—	—	6	23		—	.10	.90
	Expected		—	—	—	—	6	23	.50			
Ridge fingerlings, 1968	Observed	224	—	—	—	3	26	195		—	.07	.93
	Expected		—	—	—	1	29	194				
	Expected[c]		—	—	—	2	41	180	<.01			
Edray yearlings, 1968	Observed	68	—	—	—	1	16	51		—	.13	.87
	Expected		—	—	—	1	15	52	.70			
U. S. Fish and Wildlife Service:												
Coleville Indian yearlings, 1968	Observed	50	—	—	6	9	15	20		.06	.33	.61
	Expected		—	2	4	6	20	19	>.05			
Ontario:												
Normandale (Port Arthur) yearlings, 1966	Observed	15	—	—	—	4	8	3		—	.53	.47
	Expected		—	—	—	4	8	3	>.70			

[a] Expected values calculated on basis of allele frequencies in 1966 breeders.
[b] Expected values calculated on basis of allele frequencies in 1966 yearlings.
[c] Expected values calculated on basis of allele frequencies in 1967 breeders.

22

from the Hardy-Weinberg equilibrium tests, particularly for the LDH-B locus. At any rate, the homogeneity of intra-hatchery year class samples permitted the pooling of these samples for each hatchery in determining the significance of differences between hatchery strains.

The inter-hatchery heterogeneities for allele frequency differences are statistically significant for one, the other, or both loci in every comparison made. The heterogeneity Chi-square values, together with an indication of their level of significance, are presented in Table 6. Therefore, every hatchery brook trout strain is distinguishable from every other one on the basis of these tests of heterogeneity. The efficacy of this method of distinguishing these populations depends on the high degree of polymorphism with a relatively wide array of frequencies of different alleles extant at the two loci.

Several strain comparisons are of particular interest. The Edray, and Ridge, West Virginia, strains have established significantly different allele frequencies for both the LDH-B and the Tf loci, even though they came from the same parental stock and had been maintained as separate breeding stocks for only three generations. On the other hand, the Reynoldsdale and the Bellefonte, Pennsylvania populations are not significantly different in allele frequencies at the Tf locus for which the only comparison is possible. At the time the Bellefonte hatchery population was sampled in 1963, fish were transferred commonly from the Reynoldsdale to the Bellefonte hatchery; however, the Reynoldsdale hatchery has not received fish from Bellefonte for at least three generations.

The Pennsylvania strain and that from the Coleville Indian Reservation are remarkably similar at the transferrin locus. Indeed, these are the only two strains examined which have the Tf A allele. Although the heterogeneity between the two strains for LDH-B alleles is very highly significant, this is because there

TABLE 5.—*Heterogeneity tests of intra-hatchery year class populations for LDH-B and transferrin alleles*

Hatchery	LDH-B alleles			Transferrin alleles		
	χ^2	d.f.	P	χ^2	d.f.	P
Reynoldsdale, Pennsylvania a*	0.554	2	>.70	2.515	4	>.50
Reynoldsdale, Pennsylvania b†	2.462	2	>.20	0.351	1	>.50
Warren, New Hampshire	3.648	2	>.10	0.753	1	>.30
Ridge, West Virginia	8.838	2	>.10			

* a. Test performed using B alleles versus combined total of B′ and B″ alleles.
† b. Test performed using all 3 alleles in the two 1968 samples.

TABLE 6.—Chi-square values for heterogeneity in LDH-B and transferrin allele frequencies among hatchery populations. (n.s. = non significance; *, **, and *** represent significance of the heterogeneity at the .05, .01, and .001 levels, respectively. Degrees of freedom for each comparison are presented in parentheses.)

Hatchery population	Reynoldsdale, Pennsylvania	Warren, New Hampshire	Ridge, West Virginia	Edray, West Virginia	U. S. Fish and Wildlife Service (Coleville)	Normandale, Ontario
			Transferrin alleles			
Bellefonte, Pennsylvania	3.01 n.s. (2)	26.11*** (2)	66.91*** (2)	15.62*** (2)	2.38 n.s. (2)	10.78** (2)
Reynoldsdale, Pennsylvania	—	20.09*** (2)	146.58*** (2)	30.25*** (2)	0.19 n.s. (2)	7.12* (2)
Warren, New Hampshire	289.52*** (2)	—	101.74*** (1)	21.75*** (1)	23.58*** (2)	4.42* (1)
LDH Ridge, West Virginia	314.07*** (2)	22.91*** (2)	—	4.35* (1)	86.12*** (2)	65.70*** (1)
alleles Edray, West Virginia	402.11*** (2)	29.56*** (2)	31.07*** (2)	—	23.63*** (2)	24.29*** (1)
U. S. Fish and Wildlife Service (Coleville)	26.98*** (2)	17.55*** (1)	26.71*** (2)	43.16*** (2)	—	5.14 n.s. (2)
Normandale, Ontario	67.88*** (2)	<.01 n.s. (1)	1.53 n.s. (2)	2.82 n.s. (2)	6.66** (1)	—
Rome, New York	842.69*** (2)	232.91*** (2)	226.83*** (2)	113.56*** (2)	166.07*** (2)	28.48*** (2)

are no B″ alleles in the sample of the Coleville strain. On the other hand, the frequencies of the wild type alleles are more similar in the two than between them and any other strains. If more extensive sampling should uncover B″ alleles in the Coleville strain, one would have more reason for the interesting speculation that the Coleville strain is a Pennsylvania strain which went west. No records of the source of the Coleville introductions can be found.

Phenotype and Allele Frequencies in Wild Populations

The LDH-B phenotypes and allele frequencies in populations from seven streams are presented in Table 7. While all populations except one exhibit polymorphism, it is noted that the frequencies of variant alleles are much lower than in hatchery populations. The same statement can be made for the transferrin locus for which phenotype and allele frequencies of smaller sample sizes from six of these populations are shown in Table 8. All populations show apparent Hardy-Weinberg equilibrium for both loci except for fixation of wild type alleles for both loci in Beaver Run and except for the LDH-B in the Singley Run fish. This latter exception is due to the two homozygous B′B′ types found where less than one would be expected in this small population.

The sample sizes for the LDH phenotypes among the tributary streams of Catawissa Creek reflect our visual estimates of the population sizes in these streams. There seems to be no relationship of population size and degree of polymorphism as estimated by variant allele frequencies in these small samples. However, it seems significant that the small Beaver Run population exhibits fixation for both loci. Eckroat (1969) has found this same population to be fixed for loci controlling eye lens proteins. These findings suggest that inbreeding and/or genetic drift have been operative in this population. These forces should have been operative at two levels:

26

within the total population originally isolated in this stream, and within small breeding units in pools in the stream. Perhaps the latter accounts for the lack of fixation in the other populations sampled.

Heterogeneity Tests of Wild Populations

Intra-population tests for heterogeneity of allele frequencies at both loci among the samples taken from Marsh Creek were performed. Within the two sections of this stream there were no significant differences between samples taken: 1) by sport fishing or by electrofishing; 2) in different years; 3) in different seasons; or 4) from the nine subdivisions of Section II. The last result meant that we were unable to determine anything about the breeding structure within the stream by allele frequency analyses. This was because of the low frequencies of variant alleles being distributed throughout the length of Section II. Therefore, one would need to determine polymorphisms at many such loci in order for this method to be effective.

Inter-population heterogeneity tests permitted distinguishing most, but not all, populations from each other. Heterogeneity Chi-square values were determined for allele frequencies at each locus separately and, where possible, for the frequencies of phenotypes determined for both loci combined. Without reporting the Chi-square values, the most important findings from these tests may be summarized as follows:

1. Samples from the two sections of Marsh Creek differed significantly for LDH-B allele frequencies but not for those at the Tf locus nor for combined phenotypes.

2. When the data on phenotype and allele frequencies from the two sections of Marsh Creek were pooled and compared with pooled data from all tributary streams of Catawissa Creek, significant differences were shown between the population groups for allele frequencies at each locus and for combined phenotypes.

TABLE 7.—*LDH-B phenotype and allele frequencies in wild brook trout populations*

Population		Total	Phenotypes				P of chi-square	Alleles		
			BB	BB'	BB"	B'B'		B	B'	B"
Centre County, Pennsylvania										
Marsh Creek										
Section I	Observed	210	196	8	6	—	>.95	.97	.02	.01
	Expected		196	8	6	—				
Section II	Observed	171	169	—	2	—	>.95	.99	—	.01
	Expected		169	—	2	—				
Columbia County, Pennsylvania										
Cranberry Run	Observed	51	43	8	—	—	>.80	.92	.08	—
	Expected		43	8	—	—				
Long Hollow Run	Observed	40	36	3	—	1	>.05	.94	.06	—
	Expected		35	5	—	—				
Beaver Run	Observed	25	25	—	—	—	—	1.00	—	—
	Expected		25	—	—	—				
Singley Run	Observed	21	17	2	—	2	>.02	.86	.14	—
	Expected		16	5	—	—				
Fisher Hollow Run	Observed	32	28	—	4	—	>.70	.94	—	.06
	Expected		28	—	4	—				
Montana										
Maiden Canyon	Observed	22	12	7	—	3	>.50	.70	.30	—
	Expected		11	9	—	2				
Connecticut										
Squabble Brook	Observed	9	8	1	—	—	>.95	.94	.06	—
	Expected		8	1	—	—				

3. With three exceptions, populations from the tributary streams of Catawissa Creek are distinguishable on the basis of differences in allele frequencies at one, the other, or both loci and/or in combined phenotypes. The exceptions are the Cranberry Run, Long Hollow Run and Singley Run populations which are not distinguishable from each other on the basis of any of the three criteria. It appears that even though they have been isolated for a century, all populations have not evolved characteristically different levels of polymorphism.

NEW VARIANTS

The discovery of new variants at the Tf and the five LDH loci was an important secondary objective of this study. Except for the possible isoallele of the Tf A allele cited earlier, no new Tf variants were found in any brook trout. However, three heterozygotes for a variant, A″, at the LDH-A locus was found in the Marsh Creek population, two in Section I and one in Section II. All three were homozygous wild type for their other four LDH loci. One heterozygote for a variant, C′, at the LDH-C locus was found in the Coleville Indian strain. It was also homozygous wild type for its other four LDH loci.

In Figure 3A horizontal starch gel zymograms of vitreous humor samples from the AA″ and from BB′ and BB″ heterozygotes are contrasted. The more anodal electrophoretic mobility of the A″ form can be noted to affect the pattern of isozyme banding not only in the A-B or ubiquitous series, but also in the eye series, particularly the heterotetramer formed by three C subunits and one A subunit. In Figure 3B are shown vertical starch gel zymograms of liver extracts from the AA″ brook trout and of an AA′ hybrid between lake and brook trout (splake). Since tetramers of primarily A subunits are formed in liver, the relative mobilities of the five isozymes

TABLE 8.—*Transferrin phenotype and allele frequencies in wild brook trout populations*

Population		Total	Phenotypes			P of chi-square	Alleles	
			BB	BC	CC		B	C
Centre County, Pennsylvania								
Marsh Creek								
Section I	Observed	40	1	4	35	.20	.08	.92
	Expected		—	6	34			
Section II	Observed	100	—	10	90	>.80	.05	.95
	Expected		—	10	90			
Columbia County, Pennsylvania								
Cranberry Run	Observed	23	—	14	9	>.10	.30	.70
	Expected		2	10	11			
Long Hollow Run	Observed	13	—	3	10	>.80	.12	.88
	Expected		—	3	10			
Beaver Run	Observed	12	—	—	12	—	—	1.00
	Expected		—	—	12			
Singley Run	Observed	8	—	3	5	>.80	.19	.81
	Expected		—	3	5			
Connecticut								
Squabble Brook	Observed	9	—	1	8	>.95	.05	.95
	Expected		—	1	8			

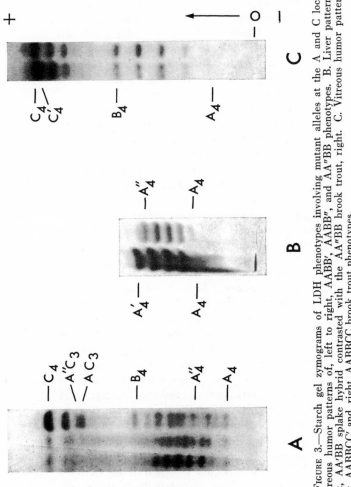

FIGURE 3.—Starch gel zymograms of LDH phenotypes involving mutant alleles at the A and C loci. A. Vitreous humor patterns of, left to right, AABB, AABB', AABB", and AA"BB phenotypes. B. Liver patterns of, left, AA'BB splake hybrid contrasted with the AA"BB brook trout, right. C. Vitreous humor pattern of, left, AABBCC' and, right, AABBCC brook trout phenotypes.

31

formed from normal A subunits and those of A′ or A″ in the two heterozygotes best distinguish the electrophoretic variants. In Figure 3C the CC′ variant heterozygote is contrasted with a normal CC homozygote in horizontal starch gel zymograms of vitreous humor samples. The probable five tetrameric isozymes of random combinations of C and C′ subunits appear as one continuous wide band in the zymogram because of the small differences in their electrophoretic mobility. Also apparent are the differences in electrophoretic mobility of the heterotetramers involving C subunits with B or A subunits.

In the course of examining zymograms of extracts of whole eyes of hundreds of fingerling and small wild brook trout, enough muscle tissue was included in the homogenates to make the D-E or muscle isozymes series visible. No electrophoretic variants were ever detected at these two loci. Nor have we ever discovered mutant forms at these loci in brook trout in our studies of thousands of fish. For that matter, neither had we ever found variants of the LDH-A and LDH-C in brook trout. Thus, while the LDH-B locus is highly polymorphic in every brook trout population, the LDH-A, -C, -D, and -E loci are practically monomorphic. Since Gillespi and Kojima (1968) have shown greater variability at those loci which control enzymes that are peripheral to the major energy-producing and anabolic pathways in *Drosophila ananassae*, it may be that the LDH-B subunit and resulting tetramers are functionally more peripheral than are the products of the other four LDH loci in brook trout. Support for this hypothesis is provided by our studies in which a null allele at the LDH-B locus is quite common in our rainbow trout populations. Rainbow trout which are homozygous for this null B allele possess only A_4 tetramers in the ubiquitous series of isozymes. On the other hand, electrophoretic variants at both the LDH-A and -C loci are quite common in rainbow trout which we have studied. Again, no variants of the LDH-D and -E loci have been found so far in rainbow trout. Of course, it should be recognized that a maximum of about one-fourth of the variants possible at a locus can be detected by electrophoresis, assuming single amino acid substitutions to be responsible for electrophoretic mobility differences (Shaw, 1965).

It seems odd that the A″ and the C′ variants were discovered in one wild and one semi-wild population, respectively, and not in any hatchery strain. One might have predicted that if a new mutant arose it would have a greater chance of being retained and leading to polymorphism in a population receiving the artificial, supposedly superior, rearing conditions in a trout hatchery. Even though only 3 mutant A″ alleles were found in 762 scored in the Marsh Creek population these occurred in three sexually mature fish in two widely separated sections of the stream and were captured during two different years. Therefore, the Marsh Creek population may be considered polymorphic for the LDH-A as well as the LDH-B loci. The Coleville Indian strain might be considered polymorphic for both the LDH-C and -B loci in that the C′ allele occurred in a sexually mature fish and in a frequency of 1%.

While examining disc gel electrophorograms for serum transferrin bands, seven other bands of serum protein components could be visualized consistently. Only one of these bands ever showed discernible electrophoretic variability and that in the Marsh Creek populations. This was a band which shows slight anodal migration to that for the transferrins and probably represents a post albumin. Although the genetic basis for the polymorphism is not yet known, fast (F), slow (S) and FS doublet bands appeared in frequencies which suggest control by codominant alleles at a single locus. If one considers each of the bands in the disc gel electrophorogram to represent a locus, then the Marsh Creek population is polymorphic for two of the eight loci for serum proteins.

TABLE 9.—*Percentages of heterozygotes for LDH-B and transferrin loci in hatchery brook trout populations*

Hatchery	LDH-B locus	Transferrin locus
Reynoldsdale, Pennsylvania	14.4	50.1
Warren, New Hampshire	46.0	49.0
Ridge, West Virginia	52.8	12.7
Edray, West Virginia	59.0	23.5
U. S. Fish and Wildlife Service (Coleville)	22.0	42.0
Normandale, Ontario	38.5	53.3
Rome, New York	64.3	–

GENERAL DISCUSSION AND SUMMARY

It was recognized that the choice of only the LDH-B and Tf loci would introduce a bias into the estimate of the total amount of genetic variability existing in brook trout. That is, polymorphism was known to exist for both loci in other fish species and, indeed, was known to exist in brook trout since our genetic analyses depended on the polymorphism within and among our inbred lines. The degree of bias could not be anticipated for it certainly was not expected that every hatchery population and all but one of the wild populations would exhibit polymorphism at both loci.

However, it is of interest to compare the amount of variability in hatchery versus wild populations. From the data shown in Tables 3, 4, 7 and 8 the actual percentage frequency of heterozygotes at each locus in each of the populations was calculated. The frequencies in hatchery populations are shown in Table 9, while those in wild populations are shown in Table 10. It is apparent that there is generally a much higher percentage of heterozygotes in hatchery than in wild brook trout. The reason for this difference is not apparent. One could speculate that inbreeding and or genetic drift are operative in isolates of the small wild populations to reduce the number of mutant alleles. This argument would require also that there be some selection for the wild type

34

alleles. Indeed, selection alone could account for the low incidence of mutant alleles. An argument against inbreeding and selection effects in a hatchery environment is the fact that of 13 inbred lines maintained for six to eight generations by single sib pair matings at the Benner Spring Station, eight lines showed segregation for LDH-B phenotypes, seven showed segregation for transferrin phenotypes, while five showed segregation at both loci. It will be interesting to examine all of these kinds of populations over the course of numbers of generations for the fate of the alleles at the two loci.

TABLE 10.—*Percentages of heterozygotes for LDH-B and transferrin loci in wild brook trout populations*

Population	LDH-B locus	Transferrin locus
Marsh Creek	4.2	10.0
Cranberry Run	15.7	60.9
Long Hollow Run	7.5	23.1
Beaver Run	0.0	0.0
Singley Run	9.5	37.5
Fisher Hollow Run	12.5	–
Maiden Canyon, Montana	31.8	–
Squabble Brook, Connecticut	11.1	11.1

If the probable polymorphism of the LDH-A and -C loci and of the possible post albumin locus are added to the two loci which we have studied in detail, then five of the assumed 13 loci discussed earlier are polymorphic in at least one population of brook trout. This becomes an estimate of 38.4% of the loci being polymorphic in the genome of this species. Although the estimate is biased by choice of loci examined and by assumptions made, it is within the range of similar percentages estimated through similar studies in other species. These percentages range from 25% in man (Harris *et al.*, 1968), to 39% in *Drosophila pseudoobscura* (Lewontin and Hubby, 1966), to 50% in the house mouse (Selander and Yang, 1969). Our estimate of 38.4% polymorphic loci is that for wild brook trout. Since only two loci were polymorphic in the domesticated hatchery strains, the value is lowered to 15.4%.

Numbers of other enzymes and proteins need to be examined in the populations we have studied as well as in other wild populations from a wider geographic range. Such a study would provide not only a better estimate of the variability in this species, but also a more efficient method of differentiating populations and of determining the breeding structure within natural populations in streams.

Although we have been unable to distinguish all wild populations through study of only two loci, we were successful in doing so for hatchery strains. As a practical matter for trout culturists, this probably means that the strains examined also differ for gene loci controlling growth and reproduction. Therefore, crosses among the various hatchery strains should show hybrid vigor for hatchery production traits. One should be particularly interested in examining the performance of the SV strain from Rome, New York, since it is a synthetic of many of the strains examined in our study and since it has been selected for some degree of disease resistance.

ACKNOWLEDGMENTS

We are grateful for the aid of the following: L. R. Eckroat, W. K. Hershberger and W. J. Morrison for technical assistance; Raymond McCreary and the staff at Benner Spring Fish Research Station for fish culture assistance; E. D. Frost for assistance in collection of the Marsh Creek populations; and the men of the state and federal agencies cited earlier for sources of the hatchery fish.

LITERATURE CITED

BARRETT, I., AND H. TSUYUKI. 1967. Serum transferrin polymorphism in some scombroid fishes. Copeia 1967(3): 551–557.

DAVIS, B. J. 1964. Disc electrophoresis II. Method and applications to human serum proteins. Ann. New York Acad. Sci. 121: 404–427.

DE LIGNY, W. 1967. Polymorphism of serum transferrins in plaice. Proc. Xth Conf. Eur. Soc. Anim. Bloodgr., (Inst. Nat. Rech. Agron., Paris, 1966) pp. 373–378.

ECKROAT, L. R. 1969. Genetic analysis in brook trout plus some interspecific comparisons of lens proteins of Salmonidae and Esocidae. Ph.D. Thesis, The Pennsylvania State University. (Unpublished).

FINE, J. M., A. DRILHON, P. AMOUCH, ET G. BOFFA. 1965. Les types de transferrines chez certains poisson migrateurs. *In* Protides of the Biological Fluids. Proc. 12th Colloq., 1964, Elsevier Publ. Co., Amsterdam, pp. 165–168.

FUJINO, K., AND T. KANG. 1968. Transferrin groups of tunas. Genetics 59: 79–91.

GILLESPI, J. H., AND K. KOJIMA. 1968. The degree of polymorphisms in enzymes involved in energy production compared to that in nonspecific enzymes in two *Drosophila ananassae* populations. Proc. Nat. Acad. Sci. 61: 582–585.

GOLDBERG, E. 1966. Lactate dehydrogenase in trout: Hybridization *in vivo* and *in vitro*. Science 151: 1091–1093.

HARRIS, H. 1966. Enzyme polymorphisms in man. Proc. Roy. Soc. (London), Ser. B, 164: 298–310.

HARRIS, H., D. A. HOPKINSON, AND J. LUFFMAN. 1968. Enzyme diversity in human populations. Ann. New York Acad. Sci. 151: 232–242.

HERSHBERGER, W. K. 1970. Physical-chemical properties of transferrin types in brook trout. Trans. Amer. Fish. Soc. 99(1): 207–218.

HOFFMAN, ANN D. 1966. Determination of transferrin types in brook trout by means of polyacrylamide disc electrophoresis. M.S. Thesis, The Pennsylvania State University. (Unpublished).

HUBBY, J. L., AND R. C. LEWONTIN. 1966. A molecular approach to the study of genic heterozygosity in natural populations. I. The number of alleles at different loci in *Drosphila pseudoobscura*. Genetics 54: 577–594.

JAMIESON, A., AND B. W. JONES. 1967. Two races of cod at Faroe. Heredity 22: 610–612.

KOEHN, R. K., AND D. I. RASMUSSEN. 1967. Polymorphic and monomorphic serum esterase heterogeneity in catostomid fish populations. Biochem. Genetics 1: 131–144.

LEWONTIN, R. C., AND J. L. HUBBY. 1966. A molecular approach to the study of genic heterozygosity in natural populations. II. Amount of variation and degree of heterozygosity in natural populations of *Drosophila pseudoobscura*. Genetics 54: 595–609.

MARKERT, C. L., AND I. FAULHABER. 1965. Lactate dehydrogenase isozyme patterns of fish. J. Exp. Zool. 159: 319–332.

MARR, J. C., AND L. M. SPRAGUE. 1963. The use of blood group characteristics in studying subpopulations of fishes. Int. Comm. for the N.W. Atlantic Fisheries, Spec. Publ. No. 4: 308–313.

MØLLER, D., AND G. NAEVDAL. 1966. Serum transferrins of some gadoid fishes. Nature 210: 317–318.

MORRISON, W. J. 1970. Nonrandom segregation of two lactate dehydrogenase subunit loci in trout. Trans. Amer. Fish. Soc. 99(1): 193–206.

MORRISON, W. J., AND J. E. WRIGHT. 1966. Genetic analysis of three lactate dehydrogenase isozyme systems in trout: Evidence for linkage of genes coding subunits A and B. J. Exp. Zool. 163: 259–270.

ODENSE, P. H., T. M. ALLEN, AND T. C. LEUNG. 1966. Multiple forms of lactate dehydrogenase and aspartate aminotransferase in herring. Canad. Jour. Biochem. 44: 1319–1326.

PARRISH, B. B. 1964. Notes on the identification of subpopulations of fish by serological and biochemical methods; the status of techniques and problems of their future application. FAO Fish. Tech. Paper No. 36, 9 pp.

PETRAS, M. L. 1967. Studies of natural populations of *Mus*. I. Biochemical polymorphisms and their bearing on breeding structure. Evolution 21: 259–274.

SELANDER, R. K., AND S. Y. YANG. 1969. Biochemical genetics and behavior in wild house mouse populations. *In* Lindzey, G., and D. D. Thiessen (eds.), Contributions to Behavior—genetic analysis: the mouse as a prototype. Appleton-Century-Crofts, New York. (*In press*).

SHAW, C. R. 1965. Electrophoretic variation in enzymes. Science 149: 936–943.

WRIGHT, J. E., AND L. M. ATHERTON. 1968. Genetic control of interallelic recombination at the LDH-B locus in brook trout. Genetics 60: 240 (Abstr.).

Polymorphism in the Esterases of Atlantic Herring

George J. Ridgway, Stuart W. Sherburne, and Robert D. Lewis

ABSTRACT

The esterase enzymes of the tissues of Atlantic herring (*Clupea harengus harengus*) were analyzed by starch gel electrophoresis. Four sets of esterase bands were distinguished by their electrophoretic mobility, their relative activity with the two substrates, alpha-naphthyl acetate and alpha-naphthyl butyrate, and their relative concentrations in plasma, liver, and heart tissues. All of the esterases were inhibited by 10^{-4} M solutions of dichlorvos, an organophosphate inhibitor, but none was inhibited by 10^{-4} M eserine sulfate or by 10^{-4} M EDTA. Polymorphism was noted in all four sets of esterases. Evidence for the genetic control of the fastest migrating set was obtained from population genetic analyses. In this set of esterases, five distinct bands occurred either singly or in pairs. The observed distribution of the three most common bands fits the hypothesis that they are controlled by a set of autosomal alleles. The two rarest bands occurred only in the heterozygous state, as would be expected. Differences in frequencies of two of the genes were detected between herring taken in Western Maine waters and on Georges Bank.

INTRODUCTION

The development of starch gel electrophoresis by Smithies (1955) and the demonstration by Hunter and Markert (1957) that electrophoretically variable enzymes could be detected directly on the starch gels by histochemical methods have greatly increased possibilities for the detection of biochemical genetic polymorphism. These advances in methodology have remarkably expanded our knowledge concerning the extent of genetic polymorphism in natural populations. The significance of this expansion of knowledge has been treated especially well in papers by Shaw (1964, 1965), Hubby and Lewontin (1966), and Lewontin and Hubby (1966).

Although electrophoretically detectable variants have been demonstrated in several enzymes of fish, this paper deals soley with esterases. Species of fishes where polymorphism in this type of enzyme have been demonstrated include freshwater roach, *Leuciscus rutilus* (Nyman, 1965); char, *Salvelinus alpinus* (Nyman, 1967); several species of catostomid fishes (Koehn and Rasmussen,

39

1967) ; several tuna species (Sprague, 1967 and Fujino and Kang, 1968) ; flounder, *Pleuronectes flesus*, and plaice, *Pleuronectes platessa* (DeLigny, 1968) ; and eel, *Anguilla anguilla* (Pantelouris and Payne, 1968). A preliminary report on esterase polymorphisms of Atlantic herring, *Clupea harengus harengus*, was presented by Naevdal and Danielsen (1967). We are studying the biochemical genetics of herring to understand their population structure in the Gulf of Maine and adjacent waters. The herring in this area are exploited by the fleets of several nations and an understanding of the population structure is essential to a program of rational management. Among the characteristics we are studying are the esterases. In this paper we illustrate the polymorphisms we have found and present population genetic data to support the hypothesis that the differences in one of the sets of esterases are genetically controlled.

MATERIALS AND METHODS

Plasmas were obtained from live herring by bleeding from the heart with a pasteur pipette into 0.5 ml of 3.8% sodium citrate and removing the red cells within 24 hours by centrifugation. Livers, skeletal muscle, and hearts were taken from either fresh or frozen herring, and extracts were made by grinding the tissue in a mortar or a small homogenizer with 10 volumes of distilled water with liver or 1 volume with heart or skeletal muscle. The extracts were clarified by centrifugation in the cold (ca 4 C). Horizontal electrophoresis was conducted with hydrolysed potato starch (Connaught Laboratories[1]) in a discontinuous buffer

---

[1] Mention of trade names in this publication does not constitute endorsement by the Bureau of Commercial Fisheries.

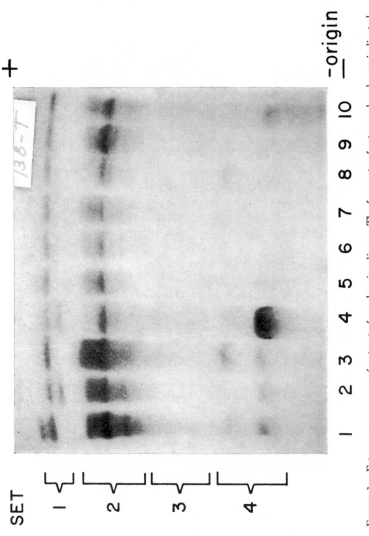

FIGURE 1.—Esterase zymogram of extracts from herring livers. The four sets of esterase bands are indicated. Phenotypes in set 1 are: sample 1-*fm*, samples 2 and 4-*ms*, samples 3 and 4-*ms*, samples 3 and 5 through 10-*mm*.

41

system modified from those used by Ferguson and Wallace (1961) and by Gahne (1966). The composition of our buffer systems was developed to provide optimal resolution of most tissue proteins of herring. The electrode buffer was 0.06 M LiOH and 0.3 M H_3BO_3. The gel buffer was prepared by adding 1 ml of the electrode buffer to each 100 ml of 0.03 M tris [tris (hydroxymethyl) aminomethane] and 0.005 M citric acid. The samples were inserted in a slit cut in the gel 5.5 cm from the cathode and on 4-mm by 6-mm pieces of Beckman #319329 filter paper. Direct current was supplied by Heathkit constant voltage power supplies. Electrophoresis was carried out at 4 C for 30 min at 165 volts. The pieces of filter paper used to apply the samples were removed and electrophoresis was continued at 320 to 350 volts for 3.5 to 4.5 hours. The gels were then sliced with a stainless steel razor band and stained for esterase activity by incubating for 30 min in a solution containing 100 ml of 0.4 M phosphate buffer (pH 6.55), 4 ml of 1% alpha naphthyl acetate ($\propto$ NA) or 1% alpha naphthyl butyrate ($\propto$ NB) in acetone, and 100 mg of Fast blue RR salt or Fast blue BB salt.

THE ESTERASE SYSTEMS OF HERRING TISSUES

Herring tissues contain at least four sets of esterase bands which can be distinguished by their electrophoretic mobility and relative activity with the two substrates used. None of the systems is inhibited by 10^{-4} M eserine sulfate or 10^{-4} M EDTA, but all are inhibited by 10^{-4} M dichlorvos (2,2, dichlorovinyl dimethyl phosphate, an organophosphate esterase inhibitor). Set 1 esterases are inhibited by 10^{-7} M dichlorvos. The esterase system moving farthest toward the anode (set 1, Figure 1) consists of one or two of several possible bands. This system stains best with $\propto$ NA, is only irregularly present in plasma, but is readily demonstrated in

42

TABLE 1.—*Frequencies of set 1 esterase phenotypes in populations of Northwest Atlantic Herring [None of the observed distributions deviates significantly from those expected from the Hardy-Weinberg law.]*

Area and number sampled	Category	Phenotypes					Gene frequencies			Chi-square 3 df
		fm	fs	mm	ms	ss	E_f	E_m	E_s	
Georges Bank N = 362	Observed	7	1	218	117	19	0.011	0.774	0.215	0.801
	Expected	6.2	1.7	216.6	120.7	16.8				
Cape May N = 96	Observed	1	0	54	37	4	0.005	0.760	0.234	0.846
	Expected	0.8	0.2	55.5	34.2	5.3				
Western Maine N = 122	Observed	2	2	61	48	9	0.016	0.705	0.279	0.994
	Expected	2.8	1.1	60.6	47.9	9.5				
Nova Scotia N = 160	Observed	2	0	82	64	12	0.006	0.719	0.275	0.872
	Expected	1.4	0.6	82.6	63.2	12.1				

extracts of liver or heart. The set 1 system is the second most intensively stained set of esterase bands with $\propto$ NA. The next set of bands (set 2, Figure 1) is the most intensively staining system with both $\propto$ NA and $\propto$ NB, and stains equally well with either substrate. The set 2 esterase system is the most complex, consisting of 2 (rarely), 3, 4, 5, 6 or 7 bands; it is very active in plasma and in liver extracts but is best resolved in extracts of heart tissue. The third set of esterase bands (set 3, Figure 1) comprises an indeterminate number of weak and diffuse bands that react most strongly in liver and plasma and with $\propto$ NB. The fourth set of esterase bands (set 4, Figure 1) consists of 2, 3, or 4 bands which stain most intensively with $\propto$ NB as substrate.

HYPOTHESES CONCERNING GENETIC CONTROL OF THE POLYMORPHISMS IN THE ESTERASES OF HERRING TISSUES

Previous studies of polymorphisms of fishes were conducted with plasma. We, however, had difficulty obtaining consistent results with plasma samples taken under field conditions. The difficulty was magnified by the much greater activity of the set 2 esterases in plasma in comparison with that of the set 1 esterases. Consequently, we report only the data on populations obtained from samples of liver or heart tissue. Five different bands, either alone or in pairs, have been found in the set 1 esterases. The most common band which has an intermediate migration, we have designated m. The next most common band (s) migrates somewhat more slowly; a third band (f) occurs less frequently and migrates faster than the m band. An even faster migrating band, encountered only once, is called f', and a band migrating more slowly than the s band has been encountered three times and is called s'. We postulate that the set 1 bands are controlled by a series of codominant autosomal alleles. Individual fish having a single band are considered homozygous

44

whereas those possessing two bands are considered heterozygous. Phenotypic frequencies obtained from four populations of herring in the Northwest Atlantic are compared with those expected on the basis of the Hardy-Weinberg law in Table 1. The remarkably close agreement provides evidence for the validity of our hypothesis with respect to the f, m, and s bands. The f' and s' bands occur too infrequently for population genetic analysis, but they have occurred only in heterozygotes, as would be expected for rare variants.

The set 2 esterases are remarkably complex. The most frequent types possess 3 or 4 bands; but types with 5, 6, and 7 bands are encountered fairly often and one fish with only 2 bands in this set has been seen. Some of the 3-, 4-, 5-, and 7-banded types are shown in Figure 2. We have attempted to develop an hypothesis which will explain the mechanism of inheritance of this system, but so far have not been successful. We have not yet attempted to develop hypotheses for the set 3 or 4 esterases since they require staining with $\propto$ NB for consistent demonstration and we have used $\propto$ NA for our routine tests.

RELATION TO PUBLISHED RESEARCH

Our findings with the esterases of herring are generally similar to those of Naevdal and Danielsen (1967) but are considerably different in detail, as would be expected, since they used a more dilute starchagar medium, a different buffer system, and were studying herring from the Northeast Atlantic rather than the Northwest Atlantic. Our set 1 systems may be the same as their Es^{m1}-Es^{m2} system but with our f type missing from their material or confused with other weak bands. The strongest group of esterases in their system contained only three bands and thus was not nearly so complicated as ours.

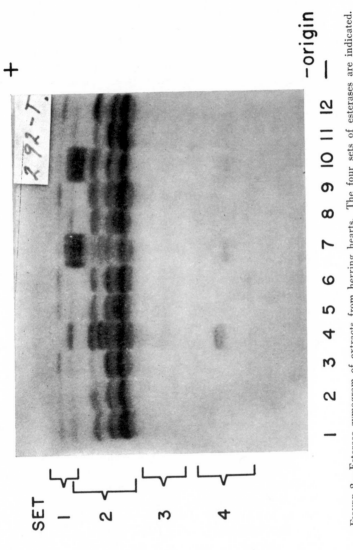

FIGURE 2.—Esterase zymogram of extracts from herring hearts. The four sets of esterases are indicated. Samples 7 and 10 contain unusual set 2 types with bands overlapping the set 1 bands.

Of the polymorphic esterases reported in other fishes only those of the eel described by Pantelouris and Payne (1968) approach the complexity of our results. The esterases of herring would all be classified as aliesterases according to their inhibition patterns; a finding in accord with those of Augustinsson (1959) on the esterases of a number of other species of fishes.

The population data obtained so far with the set 1 esterases support conclusions drawn by other authors about the population structure of herring in the Georges Bank-Gulf of Maine area. The lack of difference in the Cape May and Georges Bank samples is consistent with the findings of Zinkevich (1967). He implied from data on distribution, size, and age composition that herring found south of Long Island in the winter belong to the Georges Bank stock. The differences in the frequencies of the E_m and E_s genes between Georges Bank and Western Maine confirm the conclusion of Anthony and Boyar (1968), based on meristic data, that these subpopulations are separate.

Some of the types of set 2 esterases have been found only in samples from particular areas and may prove useful for distinction of subpopulations. Such application of the set 2 types to population problems will require additional study involving the development and testing of hypotheses regarding their mechanism of inheritance.

LITERATURE CITED

ANTHONY, V., AND H. C. BOYAR. 1968. Meristic comparison of adult herring from the Gulf of Maine and adjacent waters. Int. Comm. Northwest Atl. Fish., Res. Bull. 5: 91–98.

AUGUSTINSSON, K. B. 1959. Electrophoresis studies on blood plasma esterases. II. Avian, reptilian, amphibian and piscine plasmata. Acta. Chem. Scand. 13: 1081–1096.

DeLIGNY, W. 1968. Polymorphism of plasma esterases in flounder and plaice. Genet. Res. Cambridge 11: 179–182.

FERGUSON, K. A., AND A. L. C. WALLACE. 1961. Starch gel electrophoresis of anterior pituitary hormones. Nature (London) 190: 629–630.

FUJINO, K., AND T. KANG. 1968. Serum esterase groups of Pacific and Atlantic tunas. Copeia, 1968: 56–63.

GAHNE, B. 1966. Studies on the inheritance of electrophoretic forms of transferrins, albumins, prealbumins and plasma esterases of horses. Genetics 53: 681–694.

HUBBY, J. L., AND R. C. LEWONTIN. 1966. A molecular approach to the study of genic heterozygosity in natural populations. 1. The number of alleles at different loci in *Drosophila pseudoobscura*. Genetics 54: 577–594.

HUNTER, R. L., AND C. L. MARKERT. 1957. Histochemical demonstration of enzymes separated by zone electrophoresis in starch gels. Science (Washington) 125: 1294–1295.

KOEHN, R. K., AND D. I. RASMUSSEN. 1967. Polymorphic and monomorphic serum esterase heterogeneity in catostomid fish populations. Biochem. Genet. 1: 131–144.

LEWONTIN, R. C., AND J. L. HUBBY. 1966. A molecular approach to the study of genic heterozygosity in natural populations. II. Amount of variation and degree of heterozygosity in natural populations of *Drosophila pseudoobscura*. Genetics 54: 595–609.

NAEVDAL, G., AND D. S. DANIELSON. 1967. A preliminary report on studies of esterase phenotypes in herring. Int. Council for the Exploration of the Sea, 1967. Committee Memo H:24: 6 pp.

NYMAN, L. 1965. Inter- and intraspecific variations of proteins in fishes. Ann. Acad. Rega Sci. Upsal. 9: 1–18.

———. 1967. Protein variations in Salmonidae. Rep. Inst. Freshwater Res. Drottningholm 47: 6–38.

PANTELOURIS, E. M., AND R. H. PAYNE. 1968 Genetic variation in the eel. I. The detection of haemoglobin and esterase polymorphisms. Genet. Res. 11: 319–325.

SHAW, C. R. 1964. The use of genetic variation in the analysis of isozyme structure. *In* Subunit structure of proteins, biochemical and genetic aspects. Brookhaven Symposia in Biology Number 17: 117–130.

SHAW, C. R. 1965. Electrophoretic variation in enzymes. Science (Washington) 149: 936–943.

SMITHIES, O. 1955. Zone electrophoresis in starch gels: Group variations in the serum proteins of normal human adults. Biochem. J. 61: 629–641.

SPRAGUE, L. M. 1967. Multiple molecular forms of serum esterase in three tuna species from the Pacific Ocean. Hereditas 57: 198–204.

ZINKEVICH, V. N. 1967. Observations on the distribution of herring, *Clupea harengus* L., on Georges Bank and in adjacent waters in 1962–65. Int. Comm. Northwest Atl. Fish., Res. Bull. 4: 101–115.

Nonrandom Segregation of Two Lactate Dehydrogenase Subunit Loci in Trout[1]

WILLIAM J. MORRISON

ABSTRACT

Two of the five lactate dehydrogenase (LDH) subunit loci known to exist in salmonid fishes were obtained in heterozygous condition in the hybrid of lake trout (*Salvelinus namaycush* Walbaum) × brook trout (*S. fontinalis* Mitchill) and linkage tests were performed. These loci (and subunits), designated A and B, are both regarded as being duplicates of and therefore homologous to the B locus of higher vertebrates. In first backcrosses of doubly heterozygous males to homozygous brook trout females there was a significant deficiency of parental combinations among the offspring. However, when the doubly heterozygous parents were female, there were equal numbers of the four backcross progeny types produced, indicating independent assortment. Second backcrosses to brook trout homozygotes produced similar results. Among first backcross families there was a range of values of parental combinations of from 13.5–30.0%, while among second backcrosses this value ranged from 1.0–66.0%. The latter value is an exceptional case in which parental combinations were in excess. The data thus do not indicate linkage of the A and B genes. Possible causes of the disturbed joint segregation in males are discussed.

INTRODUCTION

With the development of the technique of gel electrophoresis it has become possible to perform genetic experiments which deal directly with gene products (proteins) rather than with superficial characters (see Shaw, 1965). Variation in the primary structure of a protein will often be reflected in a change in its electrophoretic mobility. Thus, one can study the structural gene loci for enzymes and other proteins provided that variants are available and that specific detection methods are known. Genetic variation in lactate dehydrogenase (LDH) of the deer mouse was used by Shaw and Barto (1963) to demonstrate the subunit structure of this enzyme

[1] This investigation was supported in part by grant GB 4624 to J. E. Wright from the National Science Foundation and by a National Institutes of Health Fellowship No. 4-F01-GM-34161-03 from the National Institute of General Medical Sciences.

49

which exists in mammals as a series of five tetrameric isozymes made up of two classes of polypeptide subunits, A and B.

The complex LDH isozyme makeup of trout tissues was elucidated in a similar manner by Morrison and Wright (1966). They demonstrated that trout LDH is determined by five structural gene loci through the synthesis of five subunit types, A, B, C, D and E. The A and B subunits are present in most tissues but B predominates in cardiac muscle, whereas in liver, A is most abundant as evidenced by staining intensity on zymograms. Subunits A and B combine freely as tetramers in a series of five isozymes. Another subunit, designated C, is found in eye (probably specifically in the retina), optic nerve, brain and certain muscles. Subunits D and E are found mainly in skeletal muscles in which they combine to form a series of five isozymes. Subunits A, B and C are similar in numerous properties including their *in vivo* combining affinities, heat stability, urea inhibition, and substrate inhibition, whereas D and E are similar by the same criteria (Morrison, 1969; see also Bailey and Wilson, 1968). Therefore, it is likely that subunits A and B of salmonids are not homologous to subunits A and B, respectively, of higher vertebrates as has been suggested (Morrison and Wright, 1966). Rather, A and B as well as C of trout are probably each homologous to the B subunit of mammals and birds. Similarly, the D and E subunits are probably homologous to the A subunit of higher vertebrates.

The present report deals with genetic experiments involving the loci specifying the A and B subunits in hybrids of lake trout females (*Salvelinus namaycush* Walbaum) × brook trout males (*S. fontinalis* Mitchill). The reciprocal hybrid is extremely difficult to make. These experiments are a continuation of the work of Morrison and Wright (1966), who reported a limited amount of backcross data which indicated linkage of the A and B

50

genes. The more complete analyses presented here involve reciprocal backcrosses and indicate that something other than classical linkage is responsible for the disturbed joint segregation.

MATERIALS AND METHODS

Trout LDH isozymes were examined by the same technique of starch gel electrophoresis as previously reported (Morrison and Wright, 1966) except that samples were applied to the gels in filter paper rectangles and that the gels were in a horizontal position during electrophoresis. Breeding stock were identified as to LDH type by removal of a small quantity of vitreous humor with a hypodermic syringe fitted with a Number 18 needle. Offspring were typed at the yolk sac stage or later by electrophoresis of homogenates of their heads.

This work was carried out with the use of the experimental stocks of trout maintained at the Benner Spring Fish Research Station at Bellefonte, Pennsylvania. Most of the crosses were backcrosses of doubly heterozygous splake trout to homozygous brook trout. The crosses were done reciprocally and involved both F_1 splake trout (lake trout female × brook trout male) and B_1 individuals (F_1 splake trout female × brook trout male). All lake trout were of the Cayuga Lake, New York strain and were not inbred. The brook trout used in these crosses were inbred for from three to eight generations or were F_1 hybirds between inbreds. In the tables these brook trout strains or strain crosses are indicated as ST2, ST3, ST15X0, and so forth (ST = speckled or brook trout).

RESULTS

Three forms of subunit B were found among the inbred strains of brook trout at the Benner Spring Fish Research Station, in other hatchery stocks and in certain wild populations (Figure 1). The most frequent form observed, B, was electrophoretically coincident with lake trout B. Less frequent were B′ and B″

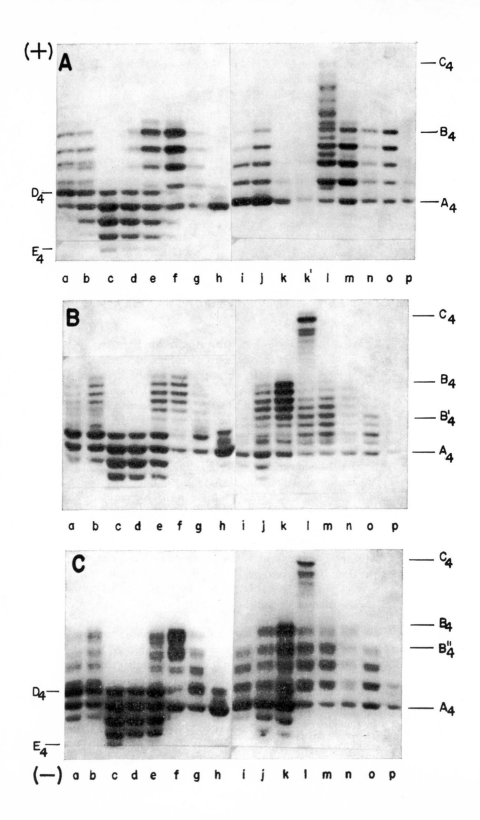

FIGURE 1.—LDH patterns of tissues of three genotypes of brook trout. A. Genotype BB, B. Genotype BB', C. Genotype BB".—a. pectoral muscle, b. mandibular muscle, c. epaxial muscle, d. adaxial muscle, e. "red" muscle, f. heart, g. stomach, h. liver, i. spleen, j. kidney, k. erythrocytes, k'. serum, l. retina and vitreous humor, m. brain, n. gill filaments, o. testis, and p. lens.

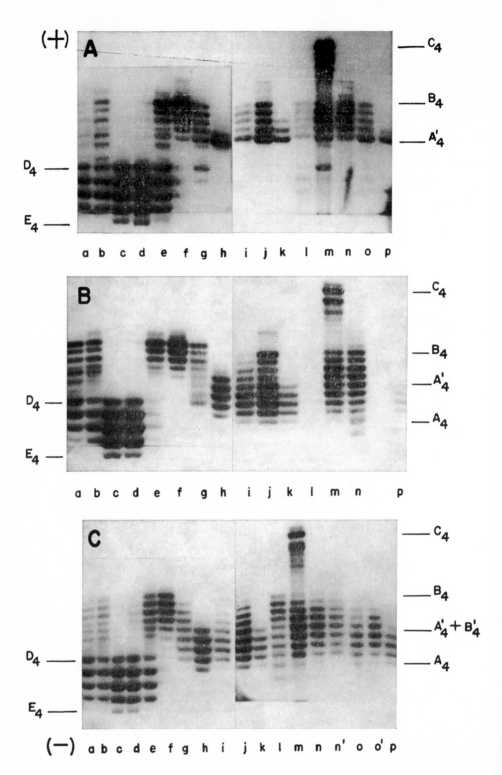

FIGURE 2.—LDH patterns of tissues of lake trout and splake. A. Lake trout, B. Splake AA'BB, and C. Splake AA'BB'.—a. pectoral muscle, b. mandibular muscle, c. epaxial muscle, d. adaxial muscle, e. red muscle, f. heart, g. stomach, h. liver, i. spleen, j. kidney, k. erythrocytes, l. serum, m. retina and vitreous humor, n. brain, n'. optic nerve, o. gill filaments, o'. testis, and p. lens.

type subunits. The fact that homopolymer B'_4 is coincident with isozyme A_2B_2 explains the apparent nine isozyme pattern of the BB' genotype which superficially resembles patterns of splake trout of genotypes AA'BB and AA'BB'. All of these peculiarities are shown diagramatically in Figure 3. Brook trout of the BB' genotype actually have the 15 isozymes that would be expected when three classes of monomers, A, B, and B', combine to form all possible tetramers. The homotetramer B_4 would be expected to be much lower in concentration in BB' than in the BB homozygote and this is realized.

The A—B system of the lake trout differs from that of the brook trout because of a difference in the A subunit (Figure 2). Lake trout A, designated A', migrates faster toward the anode than does the A subunit of brook trout. As a result, the A—B system of lake trout is a group of five isozymes appearing as a more compressed zymogram pattern than is the same system in brook trout genotype AABB. It is important to observe that isozyme A'_4 of lake trout is electrophoretically coincident with isozyme A_2B_2 of brook trout of genotype AABB and is also coincident with B'_4 of the homozygous B'B' brook trout.

There were two genotypes of F_1 splake trout dealt with, these being AA'BB and AA'BB' (Figure 2). Because the mobilities of some subunit combinations coincide, only nine bands could be detected in these genotypes but it is expected that there are 35 isozymes in AA'BB' and 15 in AA'BB. There is evidence that these numbers of isozymes do exist in these genotypes. For example, the five prominent bands seen in splake trout liver zymograms surely are the various combination of A and A'. In heart patterns of AA'BB' splake there is a decreased amount of the B_4 isozyme indicating that subunits B and B' combine with each other just as they do in brook trout genotype BB'.

The B'' subunit, another allelic form found

in the course of this study, is intermediate in mobility between subunits B and B'. This results in a third kind of five-isozyme pattern in the case of the B''B'' homozygote. In the BB'' heterozygous pattern there can be seen at least 12 bands (Figure 1-C). In the B'B'' pattern it is futile to try to observe distinct isozymes and to count them. It was found that 15 isozymes could be distinguished by artificially recombining B and B'' subunits in different proportions including 1:1, 1:2, 1:3, 2:1 and 3:1 by the salt-freeze technique (Morrison, 1969). The resulting patterns taken together reveal a certain overlap in the anodal part of the pattern which explain why only 12 isozymes are seen in the *in vivo* patterns.

As subunits A and B vary in charge, changes occur in the electrophoretic pattern of extra isozymes of the eye, brain and so forth (Figure 3). The most anodal isozyme does not change, since this is the homotetramer C_4. A sub-band always present on the anodal side of C_4 is also constant with changes in A and B. LDH homotetramers always seen to have this kind of sub-band or bands and as yet there is no explanation for them. There is evidence not to be presented here that all 15 possible tetramers of A, B and C occur in the eye. Each of the *Salvelinus* genotypes has a slightly different isozyme pattern in eye and while every band cannot be conclusively identified in each case, the pattern changes adequately demonstrate that subunits A, B and C form this series. Note especially the five bands in the eye pattern of genotype A'A'B'B' (Figure 3). This pattern is caused by the identity of net charges of A' and B' subunits and it alone is sufficient proof of the composition of this system. A few additional bands present in the A'A'B'B' pattern are not readily explainable except for the one or two bands near the most cathodal band which are probably the typical subbands found associated with most LDH homotetramers.

As was established above, the various genotypes involving combinations of normal or

57

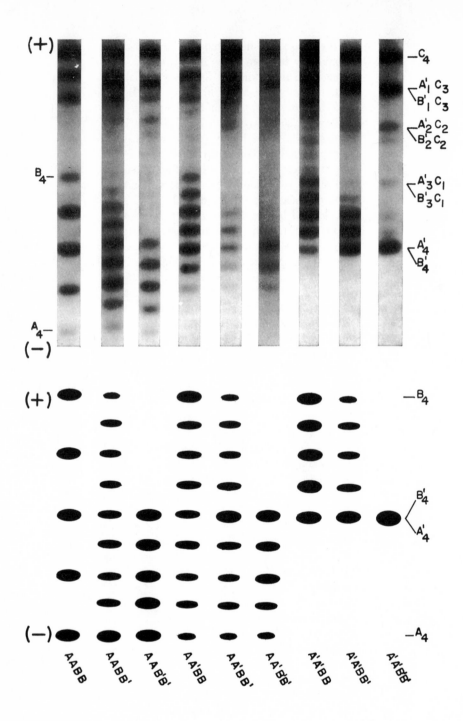

FIGURE 3.—LDH patterns of nine genotypes of F_2 splake. Above: Zymograms of eye extracts. Below: Diagrams of A-B systems.

variant alleles of genes A and B were distinguishable by their zymogram patterns. Therefore offspring types could be classified from any matings performed to establish the genetic segregation patterns (Figure 4).

F_1 splake trout were produced by crossing lake trout of genotype A'A'BB × brook of genotype AAB'B'. The schemes of mating involved in performing the first and second backcrosses are presented in Figures 5 and 6, respectively. In the first backcrosses in which the doubly heterozygous splake trout was the male parent (Table 1), disturbance of the expected 1:1:1:1 joint segregation ratio was evident in each family studied. In each family deviation as measured by Chi-square for joint segregation from the expected ratio of 1:1:1:1 for independent assortment is highly significant (P < .001). But segregation of A:A' and B:B' fits well the expected 1:1 ratio except in one family, N-127, in which .05 > P_A > .02. These are not the results which would have been expected had linkage been present; rather, the deficiency is in the *parental* classes and the excess in the *recombinant* classes. For each family the percentage parental combinations (AABB' and AA'BB) is given and they range from 13.5 to 30.0%. The data are not indicative of linkage because the parental gamete combinations AB' of brook trout and A'B of lake trout (scored as AABB' and AA'BB zymotypes in all but one family in Table 1) should be in a combined frequency of something over 50% with linkage, while gametes AB and A'B' (scored as AABB and AA'BB') should be the deficient classes. Since this was not the case, something other than classical linkage must be the cause of the disturbance of random joint segregation in the male F_1.

Table 2 contains a summary of first backcrosses in which the doubly heterozygous parents were females. In each family the ratio is indicative of independent assortment. It should be noted here that these females are sisters of the males involved in the crosses

60

TABLE 1.—*Brook trout female × F₁ splake male (first backcross)*

AABB × AA'BB'

Family	Female Strain	Male Strain	Genotypes					P³ of χ², 1 df			Percentage parental combinations
			AABB	AABB'	AA'BB	AA'BB'	Total	P_A	P_B	P_J	
N-116	ST2	F₁ Spl	36	12	13	34	95	>.9	>.3	<.001	26.3
N-120	ST2	F₁ Spl	24	4	7	37	72	>.05	>.2	<.001	15.3
N-124	ST2	F₁ Spl	81	26	19	93	219	>.7	>.1	<.001	20.5
N-125	ST2	F₁ Spl	34	10	6	23	73	>.05	>.1	<.001	21.9
N-127	ST2	F₁ Spl	36	12	5	23	76	>.02	>.1	<.001	22.4
N-128	ST2	F₁ Spl	35	8	8	26	77	>.3	>.3	<.001	20.7
N-129	ST2	F₁ Spl	38	17	16	46	117	>.3	>.3	<.001	28.2
N-130	ST2	F₁ Spl	30	4	7	39	80	>.5	>.5	<.001	13.7
N-297	ST3	F₁ Spl	39	7	8	25	79	>.1	>.05	<.001	18.9
N-299	ST3	F₁ Spl	37	17	10	26	90	>.05	>.05	<.001	30.0
N-2A	ST15X0	F₁ Spl	66	8	10	49	133	>.1	>.05	<.001	13.5

AAB'B' × AA'BB'

Family	Female Strain	Male Strain	AABB'	AAB'B'	AA'BB'	AA'B'B'	Total	P_A	P_B	P_J	
N-306	STOX12		36	11	13	40	103	.75	.9	<.001	23.3
Total			495	136	122	461	1,214				21.2

ᵃ P_A, P_B and P_J are the probabilities associated with the χ² values for segregation for A, B and for A and B jointly, each having one degree of freedom.

61

TABLE 2.—*F₁ splake female × brook trout male (first backcross)*

AA'BB' × AABB

Family	Female Strain	Male Strain	Genotypes					P of χ^2, 1 df			Percentage parental combinations
			AABB	AABB'	AA'BB	AA'BB'	Total	P_A	P_B	P_J	
N-109	F₁ Spl	ST15X0	33	40	33	34	140	>.5	>.3	>.5	52.1
B-111	F₁ Spl	ST15X0	26	36	33	25	120	>.7	>.8	>.1	57.5
N-112	F₁ Spl	ST15X0	12	18	18	24	72	>.1	>.1	>.1	50.0
N-113	F₁ Spl	ST15X0	21	28	32	30	111	>.2	>.5	>.3	54.0
Total			92	122	116	113	443	>.3	>.1	>.1	53.7

in Table 1. Thus, the second important feature of these data is that the disturbed segregation does not occur in families in which the heterozygous parents are females.

Phenotypic frequency data of the second backcrosses are summarized in Tables 3 and 4. In nearly all of the families in which the males were doubly heterozygous there is aberrant joint segregation. One family which did not exhibit a highly significant deviation from the expected 1:1:1:1 ratio for independence was family M-300 with 46% parental combinations. Little can be said about this particular family. Family M-299 is another family in which the frequencies of genotypes were statistically not significantly different from a 1:1:1:1 ratio, however there is a slightly deficiency of parental types. The same male used in this cross was used the next year in lot N-104 which indeed does have a significant deviation from random joint segregation. Families M-301 and N-133 are also families sharing the same male parent but which were spawned in two consecutive years. Their percentage parental combinations are quite similar. Family M-332 was sampled at both the sac fry stage at the age of 21 months. The percentages of parental combinations for the young and old fish were 6% and 11%, respectively.

The range of percentages of parental-type segregants in the group of second backcrosses in Tables 3 and 4 is much greater than that in the first backcrosses. This group, it should be noted, is made up of families, the male parents of which were sibs belonging to family lot K-205 (F_1 splake AA'BB brook trout, strain 14, AAB'B''). The percentages range from 1.0–66.0%. It is of interest that a value of 66% was obtained in one of these families. This is a reversal of the usual pattern since parental types make up the greater proportion of offspring in this case.

Two particularly important families are M-314 and N-110A (Figure 7). In making these crosses the same male was used, but in

(+) **A**

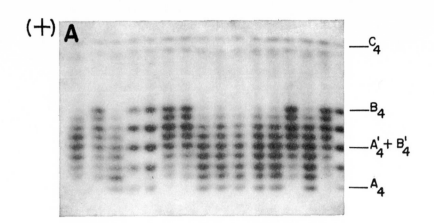

C$_4$

B$_4$

A$_4'$ + B$_4'$

A$_4$

d c b a a c c b b b b b c b c a

B

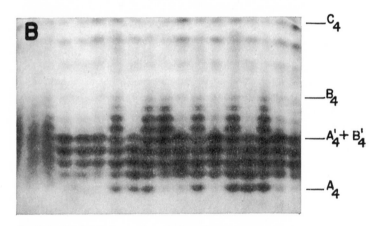

C$_4$

B$_4$

A$_4'$ + B$_4'$

A$_4$

d d d f f f b e b d f b f b e b f f

C

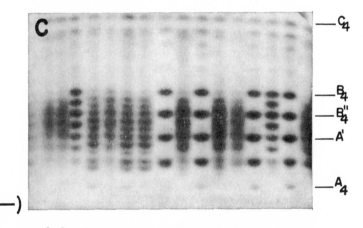

C$_4$

B$_4$

B$_4''$

A'

A$_4$

(—)

g h h c g g g g a h a h h a c a h

64

FIGURE 4.—Examples of LDH zymograms used in classifying offspring of backcrosses. Isozymes of the eye are visible since the heads of fry were used as sample material. Note that each family lot shown is segregating for four of the eight genotypes indicated in the legend above. A. Family N 104, B. Family N 110A, and C. Family N 306.—a. AABB, b. AABB′, c. AA′BB, d. AA′BB′, e. AAB′B′, f. AA′B′B′, g. AABB″, and h. AA′BB″.

65

TABLE 3.—*Brook trout female × B₁ splake male (second backcross)*

AABB × AA'BB"			Genotypes					P of x^2, 1 df			Percentage parental combinations
Family	Female Strain	Male Strain	AABB	AABB"	AA'BB	AA'BB"	Total	P_A	P_B	P_J	
M-273	ST3	SplX14	37	1	0	33	71	>.5	>.7	<.001	1.4
M-295	ST3	SplX14	19	36	35	18	108	>.8	>.1	<.005	65.7
M-297	ST3	SplX14	45	5	5	47	102	>.75	>.75	<.001	9.8
M-300	ST Ea	SplX14	22	25	24	35	106	>.2	>.1	>.3	46.2
M-299[a]	ST3	SplX14	25	19	23	28	95	>.3	>.9	>.25	44.2
N-104[a]	ST2	SplX14	34	19	19	28	100	>.5	>.5	>.02	38.0
M-500		SplX14	31	12	12	34	89	>.75	>.75	<.001	26.9
M-301[b]	ST Ea	SplX14	9	2	1	12	24	>.5	>.3	<.001	12.5
N-133[b]	ST2	SplX14	70	5	7	66	148	>.8	>.5	<.001	8.1
N-132	ST2	SplX14	22	4	3	47	76	<.01	<.005	<.001	9.2
N-134	ST2	SplX14	51	8	9	58	126	>.3	>.5	<.001	13.4
AAB'B" × AA'BB"			AABB' AAB'B' AABB" AAB'B"		AA'BB' AA'B'B' AA'BB" AA'B'B"						
M-332	STO	SplX14	55	1	6	52	114	>.8	>.3	<.001	6.1
M-332	STO	SplX14	30	2	4	20	56	>.2	>.1	<.001	10.7

[a] Same male parent.
[b] Same male parent.

TABLE 4.—*Brook trout female × B₁ splake male (second backcross)*

AABB″ × AA′BB″

Family	Female Strain	Male Strain	Genotype							P of χ², 1 df			Percentage parental combinations
			AABB	AABB″	AAB″B″	AA′BB	AA′BB″	AA′B″B″	Total	P_A	P_B	P_J	
M-330	ST 14	Spl×ST14	31	30	1	0	23	27	112	>.25	>.75	<.001	.02
		Expected	14	28	14	14	28	14					

67

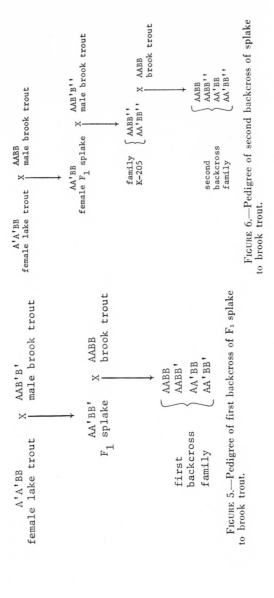

A'A'BB
female lake trout
X
AAB'B'
male brook trout
→
AA'BB'
F₁ splake
X
AABB
brook trout
→
first
backcross
family
{
AABB
AABB'
AA'BB
AA'BB'
}

FIGURE 5.—Pedigree of first backcross of F₁ splake to brook trout.

A'A'BB
female lake trout
X
AABB
male brook trout
→
AA'BB
female F₁ splake
X
AAB'B''
male brook trout
→
family
K-205
{
AABB''
AA'BB''
}
X
AABB
brook trout
→
second
backcross
family
{
AABB
AABB''
AA'BB
AA'BB''
}

FIGURE 6.—Pedigree of second backcross of splake to brook trout.

two consecutive years. The percentages of parental combinations of these families are quite similar (85 and 87%) and show that the factor which disturbs the segregation is reasonably constant from year to year. What is more significant about these lots is that they exhibit disturbed segregation in the reversed pattern, while the male that produced them was a member of a family with the usual pattern of segregation disturbance.

The data of lot M-330 recorded in Table 4 fit the pattern of the other second backcrosses. In this cross of a female brook trout AABB″ × B_1 splake trout male, AA′BB″ the estimated percentage of parental combinations is 2%.

In the recurrent backcrosses in which the doubly heterozygous parents were females (Table 5), as in similar initial backcrosses, there is good evidence of independent assortment.

An F_2 family was sampled at five months and at 21 months after spawning (Table 6). The proportions of the nine possible types in the two samples are quite comparable. Distortion of the expected 1:2:1:2:4:2:1:2:1 ratio is evident and is explainable by assuming independent assortment in the eggs and about 20% parental gene combinations in the sperm.

DISCUSSION

The results of the linkage study of the A and B loci present an interesting problem, namely that of formulating a chromosomal mechanism to explain them. The major features of the backcross data are the following: 1) Males, but not females, segregate significantly fewer parental gene combinations than nonparental combinations, 2) Among the second backcross families there is a wider range values for percentage of parental combinations than in first backcross families, and 3) In a few second backcrosses there is a reversal, which results in ratios in which parental combinations are in highest frequency.

It is worth noting that the third backcrosses to brook trout are similar to second backcrosses in their ratios (Wright, unpublished.)

69

The fact that male heterozygotes show gene segregation in one fashion and females in another is reminiscent of the widespread tendency in animals for males to show tighter linkage than females (Haldane, 1922). *Drosophila* males do not have crossing over at all. This sex difference causes the suspicion that the phenomenon causing the disturbed segregation is somehow related to crossing over. This suspicion is increased by the reversal of the deficient class from parental to nonparental in a few second backcrosses. The major question evoked by this suggestion of the involvement of crossing over is, how could parental combinations be decreased below fifty percent?

One conceivable cause of the aberrant ratios is that sperm, but not eggs, of genotypes AB′ and A′B are less viable or effective in fertilization than are sperm of genotypes AB and A′B′. However, it seems highly unlikely that parental-type sperm would be less functional than sperm of a nonparental composition with regard to LDH loci and linked factors. This gametic selection hypothesis was tested and found to be invalid (Wright, unpublished) by performing the following cross: AABB female brook trout × AA′B′B″ F_2 splake.

The ratios among progeny indicated that the A′B″ and AB′ gametes functioned equally well.

The peculiarities of salmonid cytogenetics which must be considered in seeking a model which fits the genetic data are: 1) The possibility of tetraploidy (Klose *et al.*, 1968; Ohno *et al.*, 1968), and 2) Intraindividual chromosomal rearrangements of the Robertsonian translocation type (Ohno *et al.*, 1965).

Polyploidy among the Salmonidae is not a new idea. It was first proposed by Svärdson (1945) on the basis of chromosome numbers of these fish which fall into three cytological groups with approximately 60, 80 and 100 somatic chromosomes. This suggested a polyploid series with a basic haploid number of ten. Rees (1964) measured the DNA con-

```
              AABB      X      AA'BB'

                        |
                        |
                        |
                        |
                        ↓          FAMILY    K118
                     AABB           16      (44.4%)

                     AABB'           4      (11.1%)

                     AA'BB           1      ( 2.8%)

Brook Trout   AABB    X   AA'BB'    15      (41.7%)

                        |
                        |
                        |
                        |
                        |
                        ↓       FAMILY    M314        N110A
                   AABB          12    ( 7.5%)     8   ( 8%)

                   AABB'         66    (42.5%)    50   (50%)

                   AA'BB         66    (42.5%)    37   (37%)

                   AA'BB'        12    ( 7.5%)     5   ( 5%)
```

FIGURE 7.—Pedigree of backcross families from dihybrid male heterozygotes showing reversal of parental-recombinant gamete types in nonrandom assortment.

TABLE 5.—*B₁ splake female × brook trout male (second backcross)*

AA'BB" × AABB

Family	Female Strain	Male Strain	Genotypes					P of χ², 1 df			Percentage parental combinations
			AABB	AABB"	AA'BB	AA'BB"	Total	P_A	P_B	P_J	
N-97A	SpIX14	ST15X0	24	26	22	19	91	>.3	>.9	>.5	52.7
N-228A	SpIX14	ST 3	22	21	31	19	93	>.3	>.1	>.25	55.9
M-132[a]	SpIX14	ST 8	28	24	21	35	108	.7	>.3	>.05	41.6
M-133[a]	SpIX14	ST 8	10	16	8	14	48	>.5	>.05	>.1	50.0
M-134[a]	SpIX14	ST 8	19	26	33	30	108	>.05	>.7	>.3	54.6
AA'BB" × AAB'B'			AABB'	AAB'B"	AA'BB'	AA'B'B"					
M-137	SpIX14	ST 0	11	6	9	10	36	>.7	>.5	>.3	41.6
Total			114	119	124	127	484	>.3	>.7	>.9	50.2

[a] Strain ST 8 male brook trout was male parent of each of these families.

TABLE 6.—*F₁ splake female × F₁ splake male (F₂)*

AA'BB' × AA'BB'

Family	Genotypes									
	AABB	AABB'	AAB'B'	AA'BB	AA'BB'	AA'B'B'	A'A'BB	A'A'BB'	A'A'B'B'	Total
M-108 (5 months)	18	16	3	17	23	4	0	13	13	107
M-108 (21 months)	15	14	3	23	31	10	0	6	5	107
Combined ages (observed)	33	30	6	40	54	14	0	19	18	214
Combined ages (expected)	13.4	26.7	13.4	26.7	53.5	26.7	13.4	26.7	13.4	214

tent of brown trout (*Salmo trutta*, $2N = 80$) and Atlantic salmon (*S. salar*, $2N = 60$) and found no significant difference. He concluded that fusion and fragmentation were more likely explanations for differences in the chromosomes of these species than was polyploidy.

The more recent theory of polyploidization of Ohno (1965) differs from Svärdson's in both its supporting evidence and scope. Essentially, Ohno proposed that within the order Clupeiformes most salmonid fish can be considered tetraploid in relation to clupeoid species. For example, the anchovy (*Engraulis mordax*) has 48 acrocentric chromosomes in its diploid complement while the herring (*Clupea pallasi*) has 52 chromosomes, eight of them being metacentrics, and the DNA values of these clupeoids are 44 and 28% that of mammals, respectively. Many salmonoids on the other hand have 60 to 80 chromosomes, many of which are metacentrics, and DNA values 80 to 90% that of mammals (Ohno, 1965).

Trouts and salmons, according to Ohno *et al.*, (1968), probably evolved by tetraplodization from a diploid ancestor which had 48 acrocentric chromosomes and a DNA value 40% that of mammals. Subsequent fusions of the Robertsonian type between acrocentric chromosomes, coupled with splitting of a few original acrocentrics to two smaller acrocentrics through pericentric inversions, must have produced the chromosome number of 60 to 80 and chromosome arm number of 96 to 104.

In considering tetraploidy, it must be realized that the trout are most likely not recent tetraploids if they are tetraploids at all. The inheritance patterns of the individual LDH subunits indicate normal disomic transmission. The specific patterns of distribution among tissues of the various subunits, and physiocochemical information as well, indicate that A and B, for example, are not allelic. Of course, according to the tetraploidy theory they were originally allelic. There is no known gene in trout, out of about 30 studied, which follows a pattern of tetrasomic inheritance (Wright, unpublished).

Trout chromosomes are known to form multivalents at meiosis (Ohno et al., 1965). Recent studies by Davisson (manuscript in preparation) indicate that meiotic multivalents occur in splake trout, brook trout and lake trout and there appear to be certain differences between the numbers of multivalents formed in meiosis in the males and in meiosis in the females. These multivalents might possibly reflect a residual amount of affinity of formerly homologous chromosomes; an alternative explanation is that they are a result of numerous translocations. It is a possibility that the chromosomes bearing the LDH loci are involved in a tetravalent with the chromosomes bearing the LDH B loci. This tetravalent could be the result of synapsis of parts of these because of certain homologies. Somehow, this might influence segregation so that the combinations A'B' and AB go to the same poles more frequently than AB' and A'B. It is difficult to propose an exact scheme because there are undoubtedly special mechanisms involved in meiosis of these fish, since, in spite of numerous multivalents there is no evidence of decreased fertility.

Such a special mechanism was proposed by Hamerton (1966) to occur in human interchange heterozygotes whereby preferential alternate orientation of trivalents might be the cause of the nearly complete lack of adjacent-type segregants, especially in males. These interchanges are of the Robertsonian type and involve fusion of acrocentrics. However, additional data (Hamerton, 1968) seemed to indicate that some form of prezygotic selection or meiotic drive was operative.

The second peculiarity of trout cytogenetics mentioned above was the presence of polymorphism for chromosomal rearrangements or Robertsonian translocations within the individual (Ohno et al., 1965). It is tempting to hypothesize that this phenomenon is responsible for the unusual backcross data, because a model mechanism can be readily formulated. A review of the findings of Ohno and co-workers (1965) who studied rainbow trout will demonstrate the attractiveness of this possibility.

The basic diploid complement of the rainbow trout was found to consist of 104 chromosome arms. It appeared that some elements united and dissociated in early development so that chromosome constitutions ranging from $2N = 58$ to $2N = 65$ were observed. Somatic cells of various organs tended to have a particular chromosome number. Preliminary indications are that this phenomenon also occurs in splake trout and it is significant that certain differences in chromosome numbers exist between the sexes (Davisson, unpublished). It is not difficult to imagine the LDH A and B loci occupying positions on different arms of a metacentric chromosome which disjoins at its centromere to form two telocentrics. Then fusion could occur at some point before or even during meiosis. If in doubly heterozygous splake males (AA'BB'), the arm bearing the A allele most frequently unites with the B arm and A' arm preferentially unites with the B' arm, there would be a preponderance of AB and A'B' gametes produced (nonparental combinations). The females, according to this scheme, would either lack fusion entirely or exhibit no preferential fusions, fusion being a random process in this sex.

To account for the numbers of reversals in first backcross heterozygotes in which the percentage parental combinations is in excess of 50%, a crossover (in the classical sense) could have occurred between one of the LDH loci and the centromere. The centromere region is considered here as the determining factor in the fusion process and it is assumed that the brook trout arm undergoes preferential fusion with the lake trout arm.

It is interesting to observe that Robertson

(1916) envisioned the genetic consequences of the centric breakage-fusion phenomenon. Robertson went so far as to suggest that Morgan's crossovers between the two sex linked loci in *Drosophila melanogaster*, white eye (w) and yellow body (y), could be explained on the basis of something other than chiasmata. Robertson wrote:

> . . . there is a tendency for the factors that entered together, gray body and red eye, and yellow body and white eye, to remain together. By these numbers in F_2 it may be seen that coupling was not complete but failed in the proportion of 1:84, in other words the chances are 84:1 that the factors entering together will remain together. Now, it seems possible that such a small departure from normal coupling as 1:84 might be explained by an occasional break in the V sex chromosome at the apex.

Robertson was, of course, proven to be wrong in this case. Nevertheless, the effect that breakage and fusion of chromosomes at the centromere might have on joint segregation of gene loci was recognized. As far as is known, this type of chromosomal change or exchange has never been known to produce disturbed genetic ratios.

The Robertsonian translocation model cannot be formulated in detail because of the lack of cytogenetic information. For example, are the chromosome arms fused in both lake and brook trout or are they free in both species? Might they be fused in one and not the other? At what point in development and cell cycle does the fusion occur in male splake? The work of Beçak, Beçak and Ohno (1966) with the green sunfish (*Lepomis cyanellus*) seemed to show that the specimen had to be heterozygous for a translocation at the start of its development in order to obtain a condition of intraindividual chromosomal polymorphism. They concluded that the polymorphism is a consequence of some form of somatic segregation. Whether there is some system of breakage and fusion or whether somatic segregation of chromosomes occurs in the development of splake trout testicular tissue, aberrant joint segregation of the A and B loci could result.

ACKNOWLEDGMENTS

I thank Dr. James E. Wright, Jr. for his guidance and Louisa Atherton for her cooperation and technical advice. I am grateful to Mr. Keen Buss, Mr. Raymond McCreary and the technical staff of the Benner Spring Fish Research Station who maintained the fish used in this study.

LITERATURE CITED

BAILEY, G. S., AND A. C. WILSON. 1968. Homologies between isoenzymes of fishes and those of higher vertebrates: evidence for multiple H_4 lactate dehydrogenases in trout. J. Biol. Chem. 243: 5843–5853.

BEÇAK, W., M. L. BEÇAK, AND S. OHNO. 1966. Intraindividual chromosomal polymorphism in green sunfish (*Lepomis cyanellus*) as evidence of somatic segregation. Cytogenetics 5: 313–350.

HALDANE, J. B. S. 1922. Sex ratio and unisexual sterility in hybrid animals. J. Genet. 12: 101.

HAMERTON, J. L. 1966. Chromosome segregation in three human interchanges, pp. 237–252. *In* C. D. Darlington and K. R. Lewis, Eds. Chromosomes today, Vol. 1. Oliver and Boyd, Edinburgh.

HAMERTON, J. L. 1968. Robertsonian translocations in man: evidence for prezygotic selection. Cytogenetics 7: 260–276.

KLOSE, J., U. WOLF, H. HITZEROTH, H. RITTER, N. B. ATKIN, AND S. OHNO. 1968. Duplication of the LDH gene loci by polyploidization in the fish order Clupeiformes. Humangenetik 5: 190–196.

MORRISON, W. J. 1969. Genetic and biochemical analysis of lactate dehydrogenase isozymes in trout. Ph.D. Thesis, The Pennsylvania State University. (Unpublished).

MORRISON, W. J., AND J. E. WRIGHT. 1966. Genetic analysis of three lactate dehydrogenase isozyme systems in trout: evidence for linkage of genes coding subunits A and B. J. Exp. Zool. 163: 259–270.

OHNO, S., C. STENIUS, E. FAISST, AND M. T. ZENZES. 1965. Post-zygotic chromosomal rearrangements in rainbow trout (*Salmo irideus* Gibbons). Cytogenetics 4: 117–129.

OHNO, S., U. WOLF, AND N. B. ATKIN. 1968. Evolution from fish to mammals by gene duplication. Hereditas 59: 169–187.

REES, H. 1964. The question of polyploidy in the

Salmonidae. Chromosoma 15: 275–279.

ROBERTSON, W. R. B. 1916. Chromosome studies. J. Morph. 27: 179–331.

SHAW, C. R. 1965. Electrophoretic variation in enzymes. Science 149: 937–943.

SHAW, C. R., AND E. BARTO. 1963. Genetic evidence for the subunit structure of lactate dehydrogenase isozymes. Proc. National Acad. Sci. 50: 211–214.

SVÄRDSON, G. 1945. Chromosome studies on Salmonidae. Reports from the Swedish State Institute of Fresh Water Fishery Research, Drottningholm 23: 1–151.

Some Physicochemical Properties of Transferrins in Brook Trout[1]

WILLIAM K. HERSHBERGER

ABSTRACT

Three brook trout transferrins, designated A, B and C, were purified by use of rivanol precipitation and gradient elution chromatography with DEAE-Cellulose. Physical and chemical analyses of the purified transferrins demonstrated that each type had at least four sialic acids attached to the protein moiety and that the A type of transferrin has a number of amino acid substitutions, as shown by amino acid analysis and fingerprinting. No differences could be shown between the B and C types. Analyses of molecular weights were inconclusive.

Studies on the total iron-binding capacity of the transferrins suggested a complementarity between the proteins when in heterozygous condition. The functional and genetic bases for the lethality of the AA homozygotes are discussed.

INTRODUCTION

Transferrin (siderophilin, iron-binding protein) is a beta-globulin found in the plasma of most vertebrates. This protein is highly specific for iron (Davis et al., 1962) and appears to function as a true carrier of iron (Fletcher and Huehns, 1968). However, the exact role of transferrin in iron metabolism is not completely understood.

Another characteristic of transferrin which has been extensively investigated is the high degree of polymorphism it exhibits in most species. Since the initial report demonstrating that in humans the intraspecific differences in these beta-globulins were heritable (Smithies, 1957), transferrin polymorphism has been the subject of numerous investigations. It has been shown that the inheritance for the transferrin polymorphism, where established, is due to multiple codominant autosomal alleles.

Transferrin polymorphism has been found in this laboratory in several species of trout. By electrophoretic analysis of serum on acrylamide gels, three electrophoretically dis- tinguishable types of transferrin were demonstrated in brook trout (Hoffman, 1966) and designated A, B and C in order of decreasing anodal migration. Genetic studies showed that these are determined by three codominant autosomal alleles, with the A being lethal in the homozygous condition. The designated B and C types of transferrin have very similar migration rates in polyacrylamide gels. The lethal form, A, has a much faster migration rate in the same medium, suggesting that it might be quite different structurally from the other two types.

Because of the complete selection against the homozygous AA phenotype and the apparent structural difference, it might be expected that in heterozygotes a single A allele would be at a selective disadvantage. However, subsequent genetic analysis has shown no evidence that such heterozygotes are less fit than any of the other genotypes (Wright, 1970). The A allele occurs in a relatively constant frequency in the Pennsylvania hatchery populations studied, indicating that a balanced polymorphism may be maintaining the A type of transferrin in these populations.

The apparent balanced polymorphism and the complete selection against the AA homozygotes may be related to the structure and/or function of this protein. It thus seemed desirable to purify the various types of brook trout serum transferrin in order to investigate the physical and chemical differences among them. A number of workers have purified

[1] Authorized for publication on May 5, 1969, as Paper No. 3595 in the Journal Series of the Pennsylvania Agricultural Experiment Station, University Park, in cooperation with the Benner Spring Fish Research Station, The Pennsylvania Fish Commission, Bellefonte, Pennsylvania, this work was supported in part by grant GB 4624 to J. E. Wright from the National Science Foundation and by NSF Graduate Traineeship No. 6347.

certain genetic variants of transferrin, particularly in humans, and investigated the physicochemical basis for the differences shown by electrophoresis (Roop, 1964; Wang and Sutton, 1965; Wang et al., 1966; Jeppsson, 1967). While these studies have been able to explain the electrophoretic properties, they have not been able to provide an explanation of the polymorphism, or to clarify the function. The lethality of the homozygous AA phenotype in brook trout gives us a tool by which to relate these properties in transferrin.

The objectives of the present work were: (1) to devise a purification procedure to isolate brook trout transferrin; (2) to determine the physical and chemical bases for the differences demonstrated by polyacrylamide gel electrophoresis; and (3) to attempt to determine a basis for the lethality of the homozygous AA transferrin and for the mechanism which maintains the A transferrin in a population.

EXPERIMENTAL PROCEDURE

Collection and Typing of Samples

Whole blood was obtained from selected inbred strains of brook trout maintained at the Benner Spring Fish Research Station. The blood was allowed to clot, the sample centrifuged, and the serum drawn off with a Pasteur pipette.

To determine the phenotype of the serum, all samples were typed by means of electrophoresis. The E-C discontinuous vertical acrylamide gel electrophoresis system was used for this purpose. The method employed was that given in technical bulletin #140, distributed by E-C Apparatus Corporation, Philadelphia, Pennsylvania. The gels were stained for 30 minutes with 0.2% Amido Schwartz in a solution of methanol, water and acetic acid (5:5:1, v:v:v), and then destained in an electrophoretic destainer (EC 479, E-C Apparatus Corporation) containing the same acidified methanol solution.

The transferrin types based on mobility differences were then recorded. Sera with the same phenotypes were pooled to give sufficient volume of sample for purification. Generally, the pooled samples were from siblings.

Purification of Transferrin

For purification, homozygous serum transferrin genotypes were used whenever possible. Since there were no homozygous AA genotypes available, the A transferrin was purified from sera of AC heterozygotes.

Partial purification of brook trout serum transferrins was accomplished by adding rivanol (2-ethoxy-6, 9-diaminoacridine lactate) directly to the serum (Boettcher et al., 1958), which precipitates the albumins and most of the gammaglobulins. Serum to which ferrous ammonium sulfate was added (12 μg/ml) to saturate the transferrin, was diluted with a predetermined volume (see experimental results) of 0.4% rivanol in 0.005 M Tris-HCl buffer, pH 8.8. The supernatant was filtered through raw potato starch to remove the rivanol (Sutton and Karp, 1965), and the proteins in the filtrate were adsorbed onto a 1 cm × 20 cm DEAE-Cellulose column. The proteins were eluted from the column by continuous gradient elution from 0 to 0.2 M NaCl in the 0.005 M Tris-HCl buffer.

Selected fractions were then assayed for transferrin by means of polyacrylamide disc electrophoresis. The electrophoresis procedure was that of Davis and Ornstein (Ornstein, 1964) with the Model 12 Canalco system, except that a 5% separating gel was used. Staining was accomplished with 0.7% Amido Schwartz in 7.5% acetic acid. The columns were then electrophoretically destained in 7.5% acetic acid.

Fractions with pure transferrin were pooled and dialyzed against glass double distilled water for 48 hours. The dialyzate was then lyophilized and stored at 4 C in a desiccator for future use.

Neuraminidase Treatment

Three milliliter serum samples of each of the five possible genotypes, CC, BB, BC, AC and AB, were treated with rivanol by the procedure outlined earlier. These partially purified samples were then dialyzed against double distilled water for 48 hours and lyophilized.

The lyophilized material was reconstituted with 1.5 ml of 0.1 M acetate buffer, pH 5.8, and 0.5 ml of solution was removed to use as a standard. Lyophilized neuraminidase (Sigma Chem. Co., from *Cl. perfringens*, Type V) was then dissolved in the same buffer at a concentration of two mg enzyme per ml; 0.5 ml of neuraminidase solution was added to each transferrin sample and the reaction mixture incubated at 37 C. At various time intervals 0.2 ml aliquots were removed and frozen at –20 C to stop the reaction.

The same procedure was used on purified samples of the three types of transferrin, the only change being that a weighed amount of material, 8 mg, was used for each reaction mixture.

Analysis of the neuraminidase treatment was done electrophoretically using the E-C discontinuous vertical acrylamide gel system. The procedure was the same as that for serum typing.

Total Amino Acid Analysis

For amino acid analyses, 4 mg of purified transferrin were hydrolyzed in 6 N HCl for 22 hours in evacuated and sealed tubes. The hydrolyzates were then assayed in a Beckman, Model 120C amino acid analyzer. The results, expressed as μmoles of amino acids were then converted to grams of residue per 100 grams of protein. With the hydrolysis time employed there is some decomposition of certain amino acids. The corrections for this as suggested by Moore and Stein (1963) for threonine, cystine, tyrosine and serine were used in all calculations.

Nineteen of the twenty common amino acids were calculated in this analysis. Tryptophan was not quantitated because extensive procedures are necessary to control degradation. However, a trace of tryptophan was noted in all of the hydrolyzed transferrin samples.

Molecular Weight Determinations

The molecular weights of brook trout serum transferrins A and C were determined by two methods: band sedimentation velocity measurement and the Archibald approach to equilibrium method. These were done in a Spinco model E analytical ultracentrifuge at

7 C. The buffer used to dissolve the protein samples was 0.1 M acetate buffer, pH 4.8.

Band sedimentation velocity measurements were made at a speed of 52,640 rpm in 12 mm cells. A bulk solution containing 1 N NaCl in acetate buffer was used to form the bands. Records were obtained by using UV optics and photographs were taken at 8 minute intervals for each cell. Measurements and calculations were done according to Vinograd *et al.* (1963).

The diffusion coefficients, which are necessary for the calculation of molecular weights from band sedimentation velocity measurements, were determined from synthetic boundary runs. These runs were done at a speed of 15,220 rpm with varying concentrations of protein. Records were obtained by using Schlieren optics and photographs were taken at 4 minute intervals. All calculations were extrapolated to zero concentration and corrected to water at 20 C.

The Archibald-type analysis was done at a speed of 10,589 rpm with varying concentrations of proteins. Schlieren optics were used for recording and photographs were taken at 32 minute intervals. Molecular weights were computed by the method of Schachman (1959) from enlarged tracings of the sedimentation equilibrium photographs.

Fingerprint Analysis

Both trypsin and chymotrypsin digestions of brook trout serum transferrins were used for two-dimensional electrophoretic-chromatographic peptide patterns, or "fingerprints." For digestion a 0.2% protein solution in 0.01 M NH$_4$OH, giving a final pH of 8.4, was denatured by heating at 100 C for 30 minutes. Phenol red was then added as a pH indicator. The protein was digested with trypsin at an enzyme: substrate ratio of 1:50 (w/w) or chymotrypsin at an enzyme: substrate ratio of 1:100 (w/w) for 5 hours at 37 C. After degradation, the samples were lyophilized.

Electrophoresis of a digested sample was run with a pyridine-acetic acid-water (25:1: 224, v/v/v, pH 6.4) buffer on Whatman 3MM paper, 46 cm × 57 cm. The sample was dissolved in electrophoresis buffer, applied

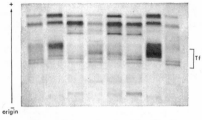

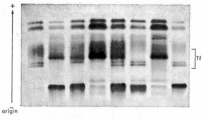

origin

Genotype   CC   BB   BC   AC   AB   BC   BB   CC

FIGURE 1.—Acrylamide gel electrophoresis of brook trout serum samples of the five non-lethal transferrins (Tf) genotypes.

origin

Genotype   CC   AA   AB   AA   AC   CC   AA   CC

FIGURE 2.—Acrylamide gel electrophoresis of brook trout serum samples of various transferrin (Tf) genotypes, including samples of the lethal homozygous AA genotype.

to the paper and run on a water-cooled plate at 2800 V for 45 minutes. After drying, descending chromatography in a second dimension was run for 18 hours with a l-butanol: acetic acid:water (4:1:5, v/v/v) solvent (Katz et al., 1959). The fingerprints were developed by dipping in 0.25% ninhydrin in acetone and drying at room temperature. They were photographed within 24–48 hours using polaroid type 47 film.

Total Iron Binding Capacity

In an attempt to ascertain a physiological basis for the lethality and the polymorphism of the brook trout transferrins, a study was done on the total iron binding capacity of these proteins. In this investigation both partially purified and completely purified material was used. With the partially purified protein, 3 ml of serum was fractionated with rivanol by the procedure outlined earlier, dialyzed and lyophilized. The pure transferrin was weighed to make a solution containing 1 mg/ml.

The transferrin samples were dialyzed for 48 hours against 0.1 M acetate buffer, pH 5.8, containing 0.01 M EDTA (Ethylenediaminetetraacetate) to remove the iron from the iron-transferrin complex. EDTA was then removed by dialysis against the acetate buffer. A final dialysis against 0.02 M $NaNCO_3$ + 0.114 M NaCl, pH 7.5, supplied a bicarbonate ion required for iron binding and raised the pH to facilitate reformation of the iron-transferrin complex.

The iron-binding capacity was then measured spectrophotometrically. 0.01 ml aliquots of a standard iron solution (0.07 mg of $Fe++$/ml) were added to 0.5 ml protein solution diluted with 2.0 ml 0.85% NaCl. After 6 minutes the increase in density was measured at 465 mμ.

RESULTS

Collection and Typing of Samples

Before any work was done with the transferrins an aliquot of every serum sample collected was subjected to electrophoresis to determine the genotype. Results of the electrophoretic typing of sera are shown in Figures 1 and 2. The serum separation in Figure 1 illustrates the five common transferrin genotypes encountered in this work; these are CC, BB, BC, and AB and AC. Sera from only the two homozygous genotypes and the heterozygous AC genotype were used for fractionation. The proteins produced by the heterozygous BC and AB genotypes were very similar in charge, and thus not amenable to the purification procedure devised.

The gel electropherogram in Figure 2 shows three sera of the homozygous AA genotype, of which only six samples have been found in this or other studies in this laboratory. However, the fish with this genotype died before crosses could be made, so the homozygous AA is still considered genetically lethal.

Pooling of serum from the same genotypes was found to be permissible since no genetic,

82

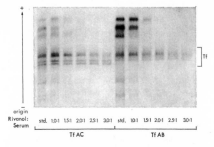

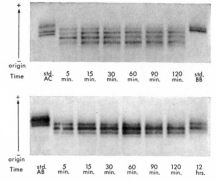

FIGURE 3.—Acrylamide gel electrophoresis of brook trout serum samples of the heterozygous AC and AB transferrin genotypes after addition of varying amounts of 0.4% rivanol solution. An untreated serum sample of each genotype is included as a standard.

FIGURE 4.—Acrylamide gel electrophoresis of partially purified transferrins after treatment for various time periods with neuraminidase. (a) An AC heterozygote with untreated samples of AC and BB phenotypes included as standards. (b) An AB heterozygote with an untreated sample of the same phenotype included as a standard.

physical or chemical evidence in the strains examined demonstrated the presence of an isoallele or a difference in the proteins produced by equivalent genes in different individuals.

Purification of Transferrin

Initially 0.4% rivanol was added at a ratio of 3:1 (v/v), rivanol to serum, as determined by Patras and Stone (1961) for the partial purification of cattle serum transferrins. However, after a few complete fractionations it was noted that the quantities of the different forms of purified transferrins were not the same. It was thus necessary to determine the volume of rivanol to serum that would give maximum purification without precipitating transferrin with the other serum proteins. The results of this experiment on heterozygous AB and AC genotypes are shown in Figure 3.

Addition of an equal volume of 0.4% rivanol did not affect the electrophoretic pattern appreciably. With a 2:1 ratio of rivanol to serum most of the albumins and gamma-globulins were precipitated, while the transferrin fractions are clearly visible and apparently unaffected by the rivanol treatment, as compared to the untreated serum. At higher concentrations of rivanol the transferrin fractions appear to diminish in concentration. The designated A form of transferrin seems to be most affected by the rivanol treatment since it decreases in amount at a lower con-

centration of rivanol than do either of the other two forms of transferrin. From these results it was concluded that the optimum ratio of rivanol to serum was 2:1, and this ratio was used in all further fractionation.

That the transferrins isolated by the methods outlined were free from contaminating proteins was shown by the Schlieren patterns obtained from the molecular weight determination and by electrophoretic analysis on polyacrylamide gels.

Physical and Chemical Analyses

Neuraminidase treatment:

The results of the electrophoretic typing of transferrin genotypes, in which two protein-staining fractions were produced by one allele, suggested that this multiple banding might be due to differing numbers of carbohydrate moieties on the polypeptide. To analyze for the carbohydrate content, transferrin preparations were treated with neuraminidase and subjected to electrophoresis. The results of this analysis on partially purified samples are shown in Figures 4 and 5.

As the time of treatment increased, the number of protein-staining fractions increased and the fractions progressively decreased in

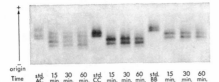

origin
Time std. 15 30 60 std. 15 30 60 std. 15 30 60
      AC min. min. min. CC min. min. min. BB min. min. min.

FIGURE 5.—Acrylamide gel electrophoresis of partially purified transferrins from an AC heterozygote, a CC homozygote and a BB homozygote after treatment for various time periods with neuraminidase. An untreated sample of each genotype is included as a standard.

anodal migration rates. Each slower migrating protein represents the loss of one more sialic acid. The results of the treatment of the sera of AC and AB heterozygotes, shown in Figures 4a and b, respectively, demonstrate a multiplicity of fractions, with a maximum of five in any one sample. Upon closer examination, the total number of fractions present for each protein was found to be five.

By examination of the CC and BB homozygote samples in Figure 5, the existence of a total of five different fractions can be more clearly seen. This indicates that there are four sialic acid residues on each protein molecule. Also the comparison of the different genotypes demonstrates that the protein moieties are determining the electrophoretic migration characteristics and not the carbohydrate portion.

Purified samples of the three forms of brook trout transferrins were also treated with neuraminidase. The results using pure protein are shown in Figures 6a, b and c. For each form of transferrin a maximum of five fractions are demonstrated up to the four hour sample and after. From the four hour sample until the last sample taken (24 hours) it appears that the sialic acid residues are being rejoined to the transferrin molecule. Close examination of the electropherograms reveals that the structure being reformed is not electrophoretically the same as that caused by the initial neuraminidase treatment. There is evidently a slight change in the shape of the molecule and perhaps a change in charge, which causes the protein to migrate differently in an acrylamide gel medium.

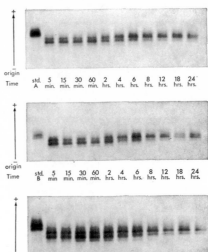

origin
Time std. 5   15   30   60   2   4   6   8   12   18   24
      A   min. min. min. min. hrs. hrs. hrs. hrs. hrs. hrs. hrs.

origin
Time std. 5   15   30   60   2   4   6   8   12   18   24
      B   min. min. min. min. hrs. hrs. hrs. hrs. hrs. hrs. hrs.

origin
Time std. 5   15   30   60   2   4   6   8   12   18   24
      C   min. min. min. min. hrs. hrs. hrs. hrs. hrs. hrs. hrs.

FIGURE 6.—Acrylamide gel electrophoresis of pure transferrins after treatment for various time periods with neuraminidase: (a) Transferrin C; (b) Transferrin B; and (c) Transferrin A. An untreated sample of each form of transferrin is included as a standard.

Total amino acid analysis:

The results of the amino acid analyses are shown in Table 1. When these are compared with the amino acid composition of human transferrin (Bezkorovainy *et al.*, 1963) a number of similarities are noted. However, a comparison of the proportional relationships of several amino acids shows some major differences also. For instance, by comparison of the total amount of basic amino acids with the total amount of acidic amino acids in human and brook trout transferrins, the brook trout protein would appear to be more acidic.

By making the same comparison between the different types of brook trout transferrin, it also becomes apparent that transferrin A probably has a larger negative charge. The aspartic acid and glutamic acid content is higher (compared to lysine, histidine and arginine) in transferrin A than in the other two transferrin types. Also, with the number of amino acid differences demonstrated in

TABLE 1.—*Amino acid composition of brook trout serum transferrins C, B and A expressed as grams of residue per 100 grams of protein*

Amino acid	Tf C	Tf B	Tf A
Lysine	10.13	8.60	8.48
Histidine	2.78	2.24	3.30
Arginine	4.74	3.96	4.25
Aspartic Acid	10.39	10.08	10.64
Threonine	6.04	5.79	5.99
Serine	6.30	6.08	6.52
Glutamic Acid	11.48	11.24	11.71
Proline	3.94	3.73	4.77
Glycine	5.22	5.11	5.39
Alanine	7.33	7.14	6.98
Half Cystine[a]	191	1.83	1.80
Valine	5.54	5.43	5.50
Methionine	1.80	1.87	2.06
Isoleucine	4.66	4.46	4.65
Leucine	7.41	7.16	7.50
Tyrosine	4.94	4.96	4.70
Phenylalanine	5.65	5.60	5.33
Tryptophan[b]	Trace	Trace	Trace

[a] Sum of cysteine and ½ cystine residues.
[b] Not quantitated (see text for reason).

TABLE 2.—*Physical constants of brook trout transferrins A and C*

	Tf A	Tf C
Diffusion coefficient ($D^\circ_{20,w}$)	10.6×10^{-7}	12.2×10^{-7}
Sedimentation coefficient ($S^\circ_{20,w}$)	3.23S	3.25S
Molecular weight (from $S^\circ_{20,w}$ and $D^\circ_{20,w}$)	29,100	25,500

transferrin A, other physical characteristics such as crystal structure, solubility and iso-electric point might be expected to be different from the other two types of transferrin.

Molecular weight determination:

The physical constants derived from the band sedimentation velocity and synthetic boundary measurements are summarized in Table 2. For the calculation of the sedimentation constants the partial specific volume was assumed to be similar to that reported for human transferrin, 0.725 cm³ per gram (Charlwood, 1963). The sedimentation constants calculated for Tf A and Tf C are somewhat lower than the values of 5.46S reported for human transferrin (Bezkorovainy et al., 1963) and 5.1S reported for rat transferrin (Charlwood, 1963).

A combination of the diffusion constant, of the partial specific volume and of the sedimentation constant gave a tentative molecular weight of 29,100 for Tf A and 25,500 for Tf C.

The Archibald-type analysis gave molecular weights for the various concentrations of Tf A of about 110,000 to 120,000 and molecular weights of 110,000 to 140,000 for the various concentrations of Tf C. These values are much higher than those obtained with the band sedimentation velocity measurements and would appear to indicate a breakdown of the protein in the band sedimentation velocity runs, perhaps into subunits. Also, the Archibald-type measurements do not indicate a

difference between the molecular weights of the two types of transferrin, as was apparently demonstrated by the band sedimentation velocity analysis.

Fingerprinting:

Both tryptic and chymotryptic digests were used for peptide mapping. Fingerprints of tryptic digests of transferrins A, B and C are shown in Figure 7. The peptide maps of transferrins B and C show some differences, none of which was consistent for all separate tryptic digests of these two forms of transferrins. These differences could be attributed to various degrees of sulfhydryl exchange or to differing degrees of oxidation among the different digests.

Compared to results with Tf B and Tf C, the fingerprint of the tryptic digest of Tf A demonstrates several apparent peptide differences. There are several peptides, marked by numbers, located in the large neutral band which are not apparent in either Tf B or Tf C. These results indicate, in conjunction with acrylamide gel electrophoresis, that the A form of transferrin is quite different from the B and C. Also, the major difference shown by

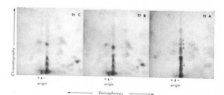

FIGURE 7.—Fingerprints of tryptic digests of purified transferrins C, B and A with the peptides outlined. The peptides marked with numbers in the fingerprint of Tf A are those which are different from those of Tf B and Tf C.

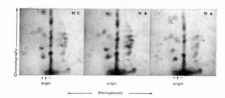

FIGURE 8.—Fingerprints of chymotryptic digests of purified transferrins C, B and A with the peptides outlined. The peptides marked with numbers in the fingerprint of Tf A are those which are different from those of Tf B and Tf C.

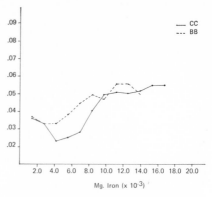

FIGURE 9.—Optical density of partially purified transferrin samples of BB and CC homozygotes after addition of fixed increments of ferrous iron solution.

electrophoresis appears to be caused by a number of amino acid substitutions, as was indicated by total amino acid analysis.

That there are a number of amino acid substitutions causing several peptide differences in the A form of transferrin is also demonstrated by the fingerprints of chymotryptic digests shown in Figure 8. By comparing the peptide map of Tf A with those of Tf B and Tf C, several peptides with differing mobilities are noted.

As with the tryptic digests, there were no consistent differences in peptides shown between the fingerprints of chymotryptic digests of Tf B and Tf C. This would indicate that the substitution causing the difference between these two proteins is "hidden" in a large peptide and/or very susceptible to chemical changes, is in a small peptide which is not detected by the methods used, or is not seen due to overlapping of other peptides.

To locate the variant peptide between Tf B and Tf C, several combinations of electrophoresis buffers, chromatographic solvents and treatment times were used. However, the results obtained from these manipulations did not demonstrate any differences. As a consequence, no variant peptide was located which would account for the apparent physicochemical difference between Tf B and Tf C.

Total iron binding capacity:

Since the major function of transferrin appears to be the absorption and subsequent distribution of iron in the body, it might be suspected that the lethality of the homozygous AA and the apparent balanced polymorphism present in the brook trout transferrins would

involve some aspect of iron transport. Therefore, a study was undertaken to compare the total iron binding capacity of the rivanol-treated serum of various genotypes and of the purified transferrins.

The results of this analysis with the homozygous BB and CC genotypes are shown in Figure 9. The volumes of the protein solutions were adjusted to give an amount of transferrin equivalent to that in one milliliter of serum. Thus from the graphs, it appears that the transferrin in one milliliter of serum is capable of binding from 10 to 13 μg of iron as a maximum. With the two homozygous genotypes, it is also apparent that not all of the iron was removed by the methods used.

However, as shown in Figure 10, the AB, AC and BC transferrins with the same treatment as the BB and CC samples do not have as much iron bound initially. Thus it would appear that the transferrins of heterozygotes lose their iron more easily. Also, the slope of the binding curve is greater with approximately the same maximum of iron bound as that of samples from homozygotes. This may also indicate that the binding is more rapid with sera of heterozygotes, but a time course study would have to be done to determine speed of binding.

The behavior of the transferrins of heterozygotes becomes even more interesting when

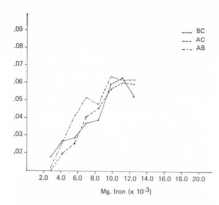

FIGURE 10.—Optical density of partially purified transferrin samples of BC, AC and AB heterozygotes after addition of fixed increments of ferrous iron solution.

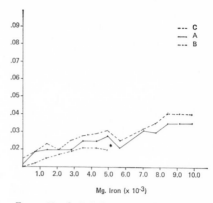

FIGURE 11.—Optical density of pure samples of transferrins C, A and B after addition of fixed increments of ferrous iron solution. *Samples lost after one measurement.

compared with the iron binding data of pure samples as shown in Figure 11. From these graphs it is apparent that the pure transferrins, A, B and C, do not differ significantly in their total iron binding capacities, in the initial amounts of bound iron, nor in the slopes of the binding curves. These results would indicate that the transferrins in combination in a heterozygous individual are capable of releasing iron more easily and may bind iron more rapidly.

DISCUSSION

Purification Procedure

Since little work has been done on purifying fish serum proteins, the procedures for purification of brook trout transferrins had to be devised from methods reported for other species. Transferrin has been purified by fractionating serum with ethanol or ammonium sulfate (Koechlin, 1952; Bezkorovainy *et al.*, 1963; Got *et al.*, 1967), but the rivanol method, originally devised by Boettcher *et al.* (1958) is much simpler and more rapid. The procedures for using rivanol in transferrin purification are well documented for human (Sutton and Karp, 1965) and cattle (Patras and Stone, 1961) material.

The results of rivanol fractionation on brook trout serum transferrins closely parallel those of other workers using human serum (Sutton and Karp, 1965) and cattle serum (Patras and Stone, 1961), with a few notable exceptions. A lower concentration of rivanol is required to precipitate the extraneous proteins from the brook trout serum. Also, as the concentration of rivanol is increased there appears to be a differential effect on the various forms of transferrin. This is particularly true for the A form of transferrin, since its concentration decreases more rapidly than either the B or C form with increasing rivanol concentration. The differential effect is probably due to a difference in solubility caused by a major structural difference in the protein.

By use of gradient elution chromatography after rivanol treatment, the need for a starch gel separation as is used in the purification of human transferrins (Sutton and Karp, 1965; Jeppsson, 1967) is eliminated. Also, there was no contamination of one form of transferrin with the other form in the samples from heterozygotes used, as is reported for human preparations (Sutton and Karp, 1965). However, not all types of heterozygotes were used for the present work. A modification of the procedure may be required if it should become necessary to purify the transferrins of the AB and BC heterozygotes, since the

87

two forms of protein in each appear to have a more similar net charge.

Ideally it would be desirable to purify a sample from a single individual. However, due to the sample size obtained from a single brook trout this could not be done. With the pooling of serum samples of like genotype, there was no evidence at any stage of the purification procedure that fish of the same genotype contained different forms of transferrin. An extra precaution was used in that, generally, all samples of the same genotype to be pooled were from the same inbred strain. This pooling would minimize any differences due to electrophoretically identical mutants.

Physical and Chemical Analyses

In the process of typing the brook trout sera on acrylamide gel electrophoresis, it was observed that more than one protein staining fraction was produced by one allele. This multiple banding was found for all three forms of transferrin. To determine if the two fractions were different proteins or if they were caused by differing carbohydrate moieties, as is reported for humans (Parker and Bearn, 1961) and cattle (Chen and Sutton, 1967), partially purified and completely purified brook trout transferrins were treated with neuraminidase.

The results of this treatment demonstrated that the multiple banding produced by one allele was due to a differing sialic acid content. There was thus no need to invoke heterogeneity of the polypeptide portion of the molecule to explain the transferrin pattern of a single allele. However, by comparing the results of neuraminidase treatment on different genotypes, it was shown that the products of the three alleles in brook trout were different proteins.

Furthermore, it was apparent that each form of brook trout serum transferrin contained at least four sialic acids in the carbohydrate portion. This is the same number that has been found in human transferrin (Parker and Bearn, 1961; Robinson and Pierce, 1964) but differs in amount from that found in swine transferrin with two sialic acids (Kristjansson and Cipera, 1963) and fowl transferrins with

either one or two sialic acids (Williams, 1966 and 1968).

One apparently anomalous result of neuraminidase treatment was that after a long treatment time, four hours or more with the concentration of enzyme used, the sialic acids appeared to be rejoined to the transferrin. One would expect that an enzyme should be required for resynthesis, but it was demonstrated in pure transferrin preparations. The explanation for the spontaneous reforming of the carbohydrate moiety is yet unsolved, but from the results obtained it appears to be a non-specific resynthesis since the resulting protein-staining fractions have a slightly different mobility.

Further analysis of the protein moiety of brook trout transferrins demonstrated that the amino acid compositions of B and C were very similar, while A is quite different in a number of residues. The results correlate with the electrophoretic behavior of transferrin A in acrylamide gels and suggest that the peptide analysis should show differences for a number of peptides.

That there were a number of peptide differences in the A type of transferrin was shown by both tryptic and chymotryptic digests. Such multiple variability in transferrins within one species has not been previously reported. Thus far only a single peptide difference has been found between varieties of transferrin within a species (Roop, 1964).

The fact that no differences in peptides were found between transferrins B and C is probably due to the technique used. It would be suspected on the basis of the previous results that probably just one peptide differs in charge, similar to what has been found in humans (Roop, 1964; Wang and Sutton, 1965; Jeppsson and Sjoquist, 1966; Wang et al., 1966). However, fingerprinting conditions or the enzyme used for digestion will have to be changed to demonstrate the difference.

The question arises from the fingerprinting results as to how this number of amino acid substitutions can be explained on a genetic basis. Two possibilities occur. One is a frameshift type of mutation and the other is a duplication of part of the gene coding for the A type of transferrin. If the latter is the case, a significant increase in molecular weight would be expected. However, the molecular weight data are still inconclusive. There was an increase of about 3600 shown in the molecular weight of transferrin A compared to transferrin C, as determined by the band sedimentation velocity method. By the Archibald-type analysis this difference was not demonstrated; in fact, on an average Tf C appeared to be some what larger than Tf A.

At present it is not known which molecular weight analysis is correct, but the results of the Archibald-type analysis are closer to those reported for othe species (Bezkorovainy et al., 1963; Charlwood, 1963; Got et al., 1967). Also, the number of peptides obtained on the fingerprints indicate that the brook trout transferrins are of a molecular weight similar to that calculated from the Archibald analysis. Recent preliminary evidence obtained by gel filtration on G-100 Sephadex supports somewhat this latter type of molecular weight analysis; transferrin B and C data indicate a molecular weight value of 78,000 for both of these proteins.

Total Iron Binding Capacity

The total iron binding capacity was measured by the change in optical density which occurs when iron is added to transferrin. When this measurement was done on rivanol-treated or partially purified transferrins, there was a difference between the homozygous and heterozygous genotypes. These differences could be explained by a faster release of iron in the heterozygotes under the conditions used and a faster binding of iron by the transferrins of heterozygotes. Both of these characteristics would seem to be advantageous when the function of iron transport is considered. If an individual has the capability of more rapid uptake and distribution of iron, there would be less stress placed on this animal under adverse conditions.

With the difference shown between the homozygous and heterozygous transferrin genotypes, it might be expected that the total iron-binding capacity would differ between proteins. Iron-binding analysis on purified proteins, however, demonstrated essentially no difference between the three types of brook trout transferrin. This relationship has been reported in human transferrin variants (Turnbull and Giblett, 1961). The similarity of iron-binding between the purified transferrins, yet differences between the homozygotes and heterozygotes, indicates that there must be some complementarily involved when two are combined in heterozygous condition.

The question then arises as to the cause of the lethality of the homozygous AA genotype, since no difference is shown in iron-binding capacity. The lethality may still be involved with the physiological function of iron-binding. Structurally the protein is quite different and in vivo may not be amenable to uptake and release of iron. On the other hand, it may be unrelated to iron binding. Rather, it could be due to some structural change which impairs attachment to receptor sites on the surface of cells for delivery of iron (Fletcher and Huehns, 1968).

The protein determined by the A allele is highly aberrant chemically and physically. The fact that heterozygous individuals show no apparent disadvantage indicates that some mechanism, perhaps related to function, is maintaining the A allele in the population. The exact mechanism for this is yet to be determined.

LITERATURE CITED

BEZKOROVAINY, A., M. E. RAFELSON, JR., AND V. LIKHITE. 1963. Isolation and partial characterization of transferrin from normal human plasma. Arch. Biochem. Biophys. 103: 371.
BOETTCHER, E. W., P. KISTLER, AND Hs. NITSCHMANN. 1958. Method of isolating the B_1-metal-combining protein from human blood plasma. Nature 181: 490.
CHARLWOOD, P. A. 1963. Ultracentrifugal characteristics of human, monkey and rat transferrins. Biochem. J. 88: 394.
CHEN, S. H., AND H. E. SUTTON. 1967. Bovine transferrins: sialic acid and the complex phenotype. Genetics 56: 425.
DAVIS, B., P. SALTMAN, AND S. BENSON. 1962. The stability constants of the iron-transferrin complex. Biochem. Biophys. Res. Comm. 8: 56.
FLETCHER, J., AND E. R. HUEHNS. 1968. Function of transferrin. Nature 218: 1211.
GOT, R., J. FONT, AND Y. GOUSSAULT. 1967. Etude sur une transferrine de Selacien, Le Grande

Roussette (*Scyllium stellare*). Comp. Biochem. 23: 317.

HOFFMAN, ANNE D. 1966. Determination of transferrin types in brook trout by means of polyacrylamide disc electrophoresis. M.S. Thesis, The Pennsylvania State University. (Unpublished).

JEPPSSON, J. O. 1967. Isolation and partial characterization of three human transferrin variants. Biochem. Biophys. Acta 140: 468.

——, AND J. SJOQUIST. 1966. Structural studies on genetic variants of human transferrin. Protides of the Biological Fluids, Vol. 14, 87 pp.

KATZ, A. M., W. J. DREYER, AND C. B. ANFINSEN. 1959. Peptide separation by two-dimensional chromatography and electrophoresis. J. Biol. Chem. 234: 2897.

KOECHLIN, B. A. 1952. Preparation and properties of serum and plasma proteins. XXVIII. The beta-metal-combining protein of human plasma. J. Am. Chem. Soc. 74: 2649.

KRISTJANSSON, F. K., AND J. D. CIPERA. 1963. The effect of sialidase on pig transferrins. Canadian J. Biochem. Physiol. 41: 2523.

MOORE, S., AND W. H. STEIN. 1963. Chromatographic determination of amino acids by the use of automatic recording equipment. Methods in Enzymology, Vol. VI, 819 pp.

ORNSTEIN, L. 1964. Disc electrophoresis. I. Background and theory. Ann. New York Acad. Sci. 121: 321.

PARKER, W. C., AND A. G. BEARN. 1961. Alterations in sialic acid content in human transferrin. Science 133: 1014.

PATRAS, B., AND W. H. STONE. 1961. Partial purification of cattle serum transferrin using rivanol. Proc. Soc. Exptl. Biol. and Med. 107: 861.

ROBINSON, J. C., AND J. E. PIERCE. 1964. Studies on inherited variants of blood proteins. III. Sequential action of neuraminidase and galactose oxidase on transferrin $B_{1-2}B_2$. J. Lab. and Clin. Med. 62: 762.

ROOP, W. E. 1964. Comparison of genetic variants and species differences in transferrin. Fed. Proc. 23: 171.

SCHACHMAN, H. K. 1959. Ultracentrifugation in Biochemistry. Academic Press, New York, N. Y.

SMITHIES, O. 1957. Variations in human serum beta-globulins. Nature 180: 1482.

SUTTON, H. E., AND G. W. KARP, JR. 1965. Adsorption of rivanol by potato starch in the isolation of transferrins. Biochem. Biophys. Acta 107: 153.

TURNBULL, A., AND E. R. GIBLETT. 1961. The binding and transport of iron by transferrin variants. J. Lab. and Clin. Med. 57: 450.

VINOGRAD, J., R. BRUNER, R. KENT, AND J. WEIGLE. 1963. Band centrifugation of macromolecules and viruses in self-generating density gradients. Proc. Nat. Acad. Sci. 49: 902.

WANG, A. C., AND H. E. SUTTON. 1965. Human transferrins C and D: chemical differences in a peptide. Science 149: 435.

——, H. E. SUTTON, AND A. RIGGS. 1966. A chemical difference between human transferrins B_2 and C. Am. J. Hum. Genet. 18: 454.

WILLIAMS, J. 1966. The sites of attachment of carbohydrate to conalbumin and transferrin in the hen. Protides of the Biological Fluids, Vol. 14, 65 pp.

——. 1968. A comparison of glycopeptides from the ovotransferrin and serum transferrin of the hen. Biochem. J. 108: 57.

WRIGHT, J. E., JR. 1970. Polymorphisms for transferrin and LDH loci in brook trout populations. Trans. Amer. Fish. Soc. 99(1): 179–192.

90

Allele Frequency Analysis of Five Soluble Protein Loci in Brook Trout, *Salvelinus fontinalis* (Mitchill)

LARRY R. ECKROAT

ABSTRACT

Phenotypes for five soluble protein loci (serum transferrin, lactate dehydrogenase, and three eye lens proteins) in native brook trout, *Salvelinus fontinalis* (Mitchill), were determined by either acrylamide or starch gel electrophoresis. An allele frequency analysis involving these five loci was performed and the resultant allele frequencies were compared (chi-square of contingency) with results of a similar study conducted on samples from the same streams in 1966.

All intra-stream, inter-year heterogeneities of the serum transferrin allele frequencies were found to be statistically insignificant. Intra-stream, inter-year heterogeneities of the three lens protein loci were statistically insignificant except for one population. In this one population the dominant allele (LP$_5$) had become fixed, which seems to be the most frequent condition in natural populations. Determination of phenotypes and calculation of allele frequencies at the LDH-B locus indicated that certain heterozygous genotypes cannot be distinguished with lens protein homogenates; however, these genotypes can be discerned with the use of whole eye homogenates. When allowances were made for the discrepancy in classification, both intra-stream and inter-year heterogeneities were found to be insignificant. Thus, indications are that allele frequencies at these five loci have remained fairly stable over a 4-year period in the five brook trout populations studied.

INTRODUCTION

The technique of electrophoresis has been demonstrated to be a highly effective method of separating mixtures of proteins which exist in extracts of animal tissues (Smithies, 1955; Davis, 1964; Ornstein, 1964). When proteins and especially enzymes separated in this way are identified, it has been common experience to find that they occur in several variant forms (polymorphisms) which are often genetically controlled (Shaw, 1965). This polymorphism of proteins has been used to demonstrate genetic variation at the intraspecific level with many different tissues in a variety of organisms.

Serum transferrins have been found to be polymorphic in fish (DeLigny, 1967; Moller and Naevdal, 1966; Hoffman, 1966, Barrett and Tsuyuki, 1967; Fujino and Kang, 1968a).

91

The lactate dehydrogenase system has been found to be polymorphic in fish (Morrison and Wright, 1966; Goldberg, 1966; Ohno et al., 1967; Market and Faulhaber, 1965; Hodgins et al., 1969; Wright and Atherton, 1970). The lens proteins have been found to be polymorphic in fish (Smith, 1962, 1965 and 1971; Smith and Goldstein, 1967; Barrett and Williams, 1967, Eckroat and Wright, 1969).

Other polymorphic systems have been identified in fish and other organisms, but the above partial review of the literature should serve to illustrate that there have been many polymorphic systems identified in fish.

The vast amount of information on polymorphic proteins has arisen through the search by fishery biologists for genetic systems which can be utilized in the identification of isolated breeding units or subpopulations. This identification of breeding populations has become an important part of fishery management programs which are concerned with the exploitation of fish populations.

Most of the genetic analyses of fish populations have been directed towards the study of allele frequencies in the commercially important fishes, particularly the tunas and herring, and the species of salmonids important to the sport fisheries. Differences in the allele frequencies of an esterase system were found in subpopulations of Atlantic Herring (*Clupea harengus harengus*) (Ridgway et al., 1970), and in skipjack tuna *Katsunonus pelamis* (Fujino and Kang, 1968b). The y blood group system failed to show heterogeneity between populations when the allele frequencies were tested statistically (Fujino and Kazama, 1968). Barrett and Tsuyuki (1967) indicated that analyses of the serum transferrin allele frequencies were useful in identifying subpopulations of tuna; however, Fujino and Kang (1968a) found no heterogeneity of transferrin allele frequencies within or between geographical areas in populations of the various tunas. The usefulness of an allele frequency analysis to distinguish fish populations was discussed by Marr and

Sprague (1963) and Fujino (1970). Eckroat (1971) utilizing the allele frequencies of three lens protein loci and Wright and Atherton (1970) using the allele frequencies of alleles at the serum transferrin and LDH-B loci were able to distinguish four hatchery populations of brook trout (*Salvelinus fontinalis*); furthermore, both studies indicated that with the above mentioned loci some but not all natural populations could be distinguished.

One of the main objectives of the 1970 study was to calculate the allele frequencies at the serum transferrin locus, the LDH-B locus and the three eye lens protein loci and compare these with the allele frequencies of a previous 1966 sample to see whether or not the allele frequencies had changed significantly over the 4-year period. Stability of allele frequencies is a necessary prerequisite to the utilization of an allele frequency analysis as a method of subpopulation identification.

MATERIALS AND METHODS

A total of 198 brook trout (*Salvelinus fontinalis*, Mitchill) were sampled from five natural populations in Columbia County, Pennsylvania. These were the same populations sampled in 1966 by Wright and Atherton (1970) and Eckroat (1971) (location presented by figure). The fish were taken by electrofishing techniques. All fish sampled were killed and collections made of eye lens. whole eye and blood. These tissues were then utilized in the present study.

The lenses were removed from a crescent-shaped cut in the cornea with a dissecting knife and a pair of dissecting forceps. Lenses were placed in a 6 × 50 mm tube under refrigeration and used for analysis immediately or after storage at –20 C. Lens samples were homogenized in glass distilled water with a ground glass homogenizer. Samples were then centrifuged for 10 minutes at 1000 g and the supernatant was utilized to determine lens protein phenotypes with the technique of acrylamide gel disc electrophoresis. The apparatus used for electrophoresis was constructed from 6 mm

93

plexiglas. Holes were arranged in a circle so that 24 gels could be placed equidistant from a platinum electrode. Procedures were similar to those of Ornstein (1964) and Davis (1964). More detailed experimental procedures are presented by Eckroat (1971).

Transferrin phenotypes were determined by acrylamide gel disc electrophoresis with the apparatus described above with a gel concentration of 7%. The tail was clipped and blood collected from the caudal artery into hematocrit tubes. The tubes were then centrifuged and cut off at the interface of the serum. The serum was then used immediately or after storage at –20 C.

LDH phenotypes were determined by horizontal starch gel electrophoresis from eye homogenates applied to filter paper wicks. The procedure has been described by Morrison (1970). After removal of the lens as described previously the whole eye was dissected free and homogenized in glass distilled water in a teflon homogenizer. The homogenate was then utilized for electrophoresis.

Eye Lens Proteins

When the eye lens proteins of brook trout are subjected to acrylamide gel disc electrophoresis, ten distinct bands are visible. Six of these ten fractions were polymorphic and genetic characterization for three of the six was made (Eckroat and Wright, 1969). Phenotypic variation of band two was expressed as varying concentration of protein; High (H); Intermediate (I); Low (L) and controlled by codominant alleles (LP_2 and lp_2). Variations of bands four and five were expressed by the presence (+) or absence (–) of the protein in the electrophorogram. The presence of band four is due to a dominant

94

TABLE 1.—*Eye lens phenotypes and allele frequencies in wild brook trout populations*

Population		Total	Band 2			P of Chi Square	LP$_2$	Band 4			Band 5		
			H	I	L			+	0	LP$_4$	+	0	LP$_5$
Cranberry Run													
1966 sample[1]	Observed	55	55	–	–	>.90	1.0	8	47	.08	–	55	1.0
	Expected		55	–	–								
1970 sample	Observed	28	28	–	–	>.90	1.0	6	22	.11	–	28	1.0
	Expected		28	–	–								
Long Hollow Run													
1966 sample[1]	Observed	59	59	–	–	>.90	1.0	24	35	.23	7	52	.66
	Expected		59	–	–								
1970 sample	Observed	55	54	1	–	>.50	.99	25	30	.26	–	55	1.0
	Expected		54	1	–								
Beaver Run													
1966 sample[1]	Observed	28	28	–	–	>.90	1.0	2	26	.04	–	28	1.0
	Expected		28	–	–								
1970 sample	Observed	41	41	–	–	>.90	1.0	–	41	0	–	41	1.0
	Expected		41	–	–								
Singley Run													
1966 sample[1]	Observed	29	29	–	–	>.90	1.0	2	27	.04	–	29	1.0
	Expected		29	–	–								
1970 sample	Observed	43	43	–	–	>.90	1.0	3	40	.04	–	43	1.0
	Expected		43	–	–								
Fisher Hollow Run													
1966 sample[1]	Observed	22	20	2	–	>.50	.96	10	12	.26	–	22	1.0
	Expected		20	2	–								
1970 sample	Observed	31	30	1	–	>.75	.99	15	16	.28	–	31	1.0
	Expected		31	–	–								

Phenotypes and dominant allele frequencies

[1] Data from Eckroat, 1971.

allele (LP_4), whereas the absence of band five is due to a dominant allele (LP_5).

The frequencies of the phenotypes and the dominant alleles for the three lens protein loci (bands 2, 4 and 5) in five natural populations of brook trout are shown in Table 1. The results of an identical study performed in 1966 (Eckroat, 1971) are also presented in Table 1 so that the two samples (1966 and 1970) can be compared. Allele frequencies were calculated on the basis of a two allele system. For the system controlled by codominant alleles, Hardy-Weinberg expectations of the three phenotypes were compared with the observed values by the chi-square test. In this regard, it can be noted that all five populations were in equilibrium. This was also true of the 1966 sample (Eckroat, 1971).

Inter-sample heterogeneities were tested by chi-square of contingency, the number of dominant alleles at each of the lens protein loci being the variables compared in the various populations. The results of these tests (Table 2) indicate that only the Long Hollow population shows a significant amount of heterogeneity. A cursory examination of the lens protein allele frequencies (Table 1) indicates that the reason for this heterogeneity is the change in the frequency of the LP_5 allele. If this allele is eliminated from the chi-square analysis, the heterogeneity of the LP_2 and LP_4 allele systems is found to be nonsignificant. The Long Hollow population was the only natural population in which the lp_5 allele was ever observed and it is also extremely rare among hatchery populations (Eckroat, 1971). It is possible that the location and nature of Long Hollow Run may have lent itself to a planting of hatchery fish by a local sportsman's club. This could account for the relatively high frequency of the lp_5 allele in the 1966 sample.

LDH Isozyme

The terminology used by Morrison and Wright (1966) to identify five loci specifying

TABLE 2.—*Heterogeneity test of intra-stream year-class populations for lens protein loci*

Population	χ^2	df	P
Cranberry Run	0.00	2	1.0
Long Hollow Run[1]	62.65	2	<.0005
Long Hollow Run[2]	0.76	1	>.25
Beaver Run	2.06	2	>.25
Singley Run	0.00	2	1.0
Fisher Hollow Run	0.14	2	>.90

[1] Test performed with lens protein alleles LP_2, LP_4, and LP_5.

[2] Test performed with only lens protein alleles LP_2 and LP_4.

five sub-units involved in the production of three series of LDH isozymes in brook trout has been followed in this study. The LDH-A and -B loci specify sub-units A and B which produce isozymes occurring in practically every tissue. This series has been termed the ubiquitous series by Wright and Atherton (1970). The LDH-C locus specifies the retinal or C sub-unit which combines with A and B to produce isozymes occurring in the eye and brain and is designated the eye series. The D and E sub-units are restricted to muscle and were not utilized in this study.

There are six possible phenotypes involving the three codominant alleles at the B locus (B, B' and B") these are all distinguishable on starch gel zymograms of eye homogenates (Morrison and Wright, 1966; Wright and Atherton, 1970). The LDH-B phenotypes and allele frequencies from the five populations sampled in this study are presented in Table 3. The results of a similar study by Wright and Atherton (1970) are presented for comparison to the present sample. The expected numbers for each phenotype were calculated with the allele frequencies.

A cursory examination of the LDH-B allele frequencies indicates that in three populations (Cranberry Run, Long Hollow Run and Singley Run) the 1966 sample indicated the presence of the B' allele but not the B" allele; the 1970 sample indicates the reverse situation. Probably this difference can be attributed to misclassification by Wright and Atherton

TABLE 3.—*LDH-B phenotype and allele frequencies in wild brook trout populations*

Population		Total	BB	BB'	BB"	B'B'	B'B"	B"B"	P of Chi Square	Alleles B	B'	B"
Cranberry Run												
1966 sample[1]	Observed	51	43	8	—	—	—	—	>.80	.92	.08	—
	Expected		43	8	—	—	—	—				
1970 sample	Observed	30	28	—	2	—	—	—	>.50	.97	—	.03
	Expected		28	—	2	—	—	—				
Long Hollow Run												
1966 sample[1]	Observed	40	36	3	—	1	—	—	>.05	.94	.06	—
	Expected		35	5	—	—	—	—				
1970 sample	Observed	57	53	—	4	—	—	—	>.75	.96	—	.04
	Expected		53	—	4	—	—	—				
Beaver Run												
1966 sample[1]	Observed	25	25	—	—	—	—	—	—	1.00	—	—
	Expected		25	—	—	—	—	—				
1970 sample	Observed	41	38	3	—	—	—	—	>.75	.96	.04	—
	Expected		38	3	—	—	—	—				
Singley Run												
1966 sample[1]	Observed	21	17	2	—	2	—	—	>.02	.86	.14	—
	Expected		16	5	—	—	—	—				
1970 sample	Observed	44	25	—	12	—	—	7	>.025	.70	—	.30
	Expected		22	—	18	—	—	4				
Fisher Hollow Run												
1966 sample[1]	Observed	32	28	—	4	—	—	—	>.70	.94	—	.06
	Expected		28	—	4	—	—	—				
1970 sample	Observed	31	25	—	6	—	—	—	>.75	.90	—	.10
	Expected		25	—	6	—	—	—				

[1] Data from Wright and Atherton (1970).

98

TABLE 4.—*Heterogeneity tests of intra-stream year-class populations for LDH-B alleles*

Population	χ^2	df	P
Cranberry Run	10.91	2	<.005
Cranberry Run[1]	1.71	1	>.10
Long Hollow	10.80	2	<.01
Long Hollow[1]	1.77	1	>.50
Beaver Run	1.53	1	>.10
Singley Run	24.74	2	<.005
Singley Run[1]	3.07	1	>.05
Fisher Hollow Run	.44	1	>.75

[1] Test performed by combining the B' and B" alleles.

because these populations were originally classified by acrylamide gel electrophoresis with eye homogenates. Later when it was discovered that all six phenotypes could not be identified on acrylamide gel (Wright and Atherton, 1970), these populations were reclassified by means of the LDH starch zymograms of the eye lens protein. This in all probability led to the misclassification because the BB' and BB" genotypes cannot be distinguished with lens protein homogenates which contain predominately LDH composed of A sub-units (Morrison, 1970). The genotypes BB, B'B', B"B" and B'B" can be distinguished with lens protein homogenates.

When inter-sample heterogeneities were calculated, only Beaver Run and Fisher Hollow showed nonsignificant heterogeneity between year samples (Table 4). Because of the above mentioned discrepancies in classification of the B' vs. B" allele frequencies, the Cranberry Run, Long Hollow Run, and Singley Run populations show significant heterogeneities between samples. If one reruns the contingency tests by testing the number of B alleles versus the combined total of the B' and B" alleles, the heterogeneities are insignificant (Table 4).

All of the populations except the Singley Run population are in Hardy-Weinberg equilibrium. It should be noted the Singley Run population was not in Hardy-Weinberg equilibrium in 1966 (Wright and Atherton, 1970).

This tends to offer strong evidence that the Singley Run population is not genetically stable in regard to the LDH-B allele. The 30% frequency of the B'' allele in this population is extremely high and is greater than has ever been observed in any natural or hatchery population. This change can possibly be accounted for by genetic drift. It should also be noted that the sample size of the Singley Run population in 1970 was considerably larger than it was in 1966.

In the Long Hollow population four C mutants appeared which appeared to be different from the C' mutant reported by Wright and Atherton (1970). This can tentatively be classified as a C'' mutant allele and occurred at approximately a 4% frequency. These were small fish and did not allow enough material for comparison with the C' mutants; further studies are being carried out in this regard.

Serum Transferrin

The genetic basis for the polymorphism at the serum transferrin locus in brook trout was determined by Hoffman (1966). Three codominant autosomal alleles control three electrophoretic variant transferrins designated as A, B and C in increasing order of anodal migration. Six phenotypes and genotypes are resolvable on acrylamide gel electrophorograms. These six types are illustrated and described by Hershberger (1970).

The transferrin phenotype and allele frequencies observed in the five populations of brook trout studied are presented in Table 5. The data of Wright and Atherton (1970) are also presented in Table 5 for comparison between the two samples. Expected values, according to the Hardy-Weinberg Law, were calculated from the allele frequencies. Chi-square tests indicate that all of the populations can be considered to be in genetic equilibrium.

Inter-sample heterogeneity tests indicated

Table 5.—*Transferrin phenotype and allele frequencies in wild brook trout populations*

Population		Total	Phenotypes			P of Chi Square	Alleles	
			BB	BC	CC		B	C
Cranberry Run								
1966 sample[1]	Observed	23	—	14	9	>.10	.30	.70
	Expected		2	10	11			
1970 sample	Observed	26	2	14	10	>.25	.35	.65
	Expected		3	12	11			
Long Hollow Run								
1966 sample[1]	Observed	13	—	3	10	>.80	.12	.88
	Expected		3	3	10			
1970 sample	Observed	24	3	6	15	>.10	.25	.75
	Expected		2	9	13			
Beaver Run								
1966 sample[1]	Observed	12	—	—	12	—	—	1.0
	Expected		1	1	12			
1970 sample	Observed	38	1	4	33	>.25	.08	.92
	Expected		—	6	32			
Singley Run								
1966 sample[1]	Observed	8	—	3	5	>.80	.19	.81
	Expected		—	3	5			
1970 sample	Observed	20	—	4	16	>.75	.10	.90
	Expected		—	4	16			
Fisher Hollow Run								
1966 sample[1]	No Data Presented							
1970 sample	Observed	30	1	13	16	>.05	.25	.75
	Expected		2	17	11			

[1] Data from Wright and Atherton (1970).

TABLE 6.—*Heterogeneity tests of intra-stream year-class populations for transferrin alleles*

Population	χ^2	df	P
Cranberry Run	.19	1	>.50
Long Hollow Run	1.50	1	>.25
Beaver Run	1.25	1	>.25
Singley Run	.80	1	>.25
Fisher Hollow Run	(Sampled only in 1970)		

that for the five natural brook trout populations there was no statistically significant difference for the transferrin allele frequencies between the 1966 sample and the 1970 sample (Table 6). It can be noted that all five populations were polymorphic at the transferrin locus. This conflicts slightly with the fact that Wright and Atherton (1970) reported that the Beaver Run population was fixed for the C allele, but their findings were based upon a sample of 12 fish.

ACKNOWLEDGMENTS

This study was supported financially by a National Science Foundation Institutional Grant to The Pennsylvania State University.

I wish to thank Dale E. Smeck and my father for assistance in the collection of samples. I express sincere appreciation to my wife, Cozella, for technical assistance. Thanks also go to J. E. Wright, Jr. for assistance in classification of LDH genotypes.

LITERATURE CITED

BARRETT, I., AND H. TSUYUKI. 1967. Serum transferrin polymorphism in some scombroid fishes. Copeia 1967 (3): 551–557.

————, AND A. WILLIAMS. 1967. Soluble lens proteins of some scombroid fishes. Copeia 1967 (2): 468–471.

DAVIS, B. J. 1964. Disc electrophoresis II. Method and applications to human serum proteins. Ann. New York Acad. Sci. 121: 404–427.

DeLIGNY, W. 1967. Polymorphism of serum transferrins in plaice. Proc. Xth conf. Eur. Soc. Anim. Bloodgr. (Inst. Nat. Rech. Agron. Paris, 1966) p. 373–378.

ECKROAT, L. R. 1971. Lens protein polymorphisms in hatchery and natural populations of brook trout *Salvelinus fontinalis* (Mitchill). Trans. Amer. Fish. Soc. 100 (3): 527–537.

————, AND J. E. WRIGHT, JR. 1969. Genetic analyses of soluble lens protein polymorphism in brook trout (*Salvelinus fontinalis*). Copeia 1969 (3): 466–473.

FUJINO, K. 1970. Immunological and biochemical genetics of tunas. Trans. Amer. Fish. Soc. 99 (1): 152–178.

————, AND T. KANG. 1968a. Transferrin groups of tunas. Genetics 59: 79–91.

————, AND ————. 1968b. Serum esterase groups of Pacific and Atlantic tunas. Copeia 1968 (1): 56–63.

————, AND T. K. KAZAMA. 1968. The Y system of skipjack tuna blood groups. Vox Sang. 14: 383–395.

GOLDBERG, E. 1966. Lactate dehydrogenase in trout: hybridization in vivo and in vitro. Science 151: 1091–1093.

HERSHBERGER, W. K. 1970. Physical-chemical properties of transferrin types in brook trout. Trans. Amer. Fish. Soc. 99(1): 207–218.

HODGINS, H. O., W. E. AMER, AND F. M. UTTER. 1969. Variants of lactate dehydrogenase isozymes in sera of sockeye salmon (*Oncorhynchus nerka*). J. Fish Res. Bd. Canada 26(1): 15–19.

HOFFMAN, A. D. 1966. Determination of transferrin types in brook trout by means of polyacrylamide disc electrophoresis. M.S. Thesis, The Pennsylvania State University (unpublished).

MARKET, C. L., AND I. FAULHABER. 1965. Lactate dehydrogenase isozyme patterns of fish. J. Exp. Zool. 159: 319–332.

MARR, J. C., AND L. M. SPRAGUE. 1963. The use of blood group characteristics in studying subpopulations of fishes. Int. Comm. N. W. Atl. Fish., Spec. Pub. No. 4: 308–313.

MOLLER, D., AND G. NAEVDAL. 1966. Serum transferrins of some ganoid fishes. Nature 210: 317–318.

MORRISON, W. J. 1970. Nonrandom segregation of two lactate dehydrogenase sub-unit loci in trout. Trans. Amer. Fish. Soc. 99(1): 193–206.

————, AND J. E. WRIGHT. 1966. Genetic analysis of three lactate dehydrogenase isozyme systems in trout: evidence for linkage of genes coding sub-units A and B. J. Exp. Zool. 163: 259–270.

ORNSTEIN, L. 1964. Disc electrophoresis I. Background and theory. Ann. New York Acad. Sci. 121: 321–349.

OHNO, S., J. KLEIN, J. POOLE, C. HARRIS, A. DESTRIE, AND M. MORRISON. 1967. Genetic control of LDH formation in a hagfish, *Eptatretus stoutii*. Science 156: 96–98.

RIDGWAY, G., S. SHERBURNE, AND R. LEWIS. 1970. Polymorphism in the esterases of Atlantic herring. Trans. Amer. Fish. Soc. 99(1): 179–192.

SHAW, C. R. 1965. Electrophoretic variation in enzymes. Science 149: 936.

SMITH, A. C. 1962. The electrophoretic characteristics of albacore, bluefin tuna and kelpbass eyelens proteins. Calif. Fish Game 48(3): 199–201.

————. 1965. Intraspecific eye-lens protein differences in yellow fin tuna, *Thunnus albacares*. Calif. Fish Game 51(3): 163–167.

————. 1971. The soluble proteins in eye lens nuclei of albacore, blue fin tuna and bonito. Comp. Biochem. Phys. 393: 719–724.

————, AND R. A. GOLDSTEIN. 1967. Variation in

protein composition of the eye lens nucleus in ocean whitefish, *Caulolatilus princeps*. Comp. Biochem. Phys. 23: 533–539.

SMITHIES, O. 1955. Zone electrophoresis in starch gel: Group variations in the serum proteins of normal human adults. Biochem. J. 61: 629–641.

WRIGHT, J. E., JR., AND L. M. ATHERTON. 1970. Polymorphism for LDH and transferrin loci in brook trout populations. Trans. Amer. Fish. Soc. 99(1): 179–192.

Centric Fusion and Trisomy for the LDH-B Locus in Brook Trout, Salvelinus fontinalis

Muriel Trask Davisson

J. E. Wright
Louisa M. Atherton

Abstract. *Cytogenetic analyses showed that a trisomic male brook trout of genotype BB'B" for one of the lactate dehydrogenase subunit loci had a karyotype with two extra arms appearing as a metacentric chromosome. The metacentric chromosome probably arose through centric fusion of two acrocentric or telocentric chromosomes—one of which carried the locus for subunit B—followed by nondisjunction.*

In the course of screening a population of brook trout for use in intragenic recombination studies (*1*) at the locus specifying the B subunit of the ubiquitous system of lactate dehydrogenase (LDH) in trout (*2*), a male showed a zymogram pattern that indicated trisomy for three *LDH-B* alleles—genotype *BB'B"*. Breeding and cytogenetic analyses of this male confirmed the trisomy and indicated that it probably arose through a spontaneous centric fusion of two acrocentric or telocentric chromosomes, followed by nondisjunction.

Starch gel electrophoresis of eye tissue (*2*) permits unambiguous detection of all genotypes for the *LDH-B* locus. This is so because gene dosage effects, as well as allelic differences, are reflected in differentially stained homo- and heterotetramers of unique electrophoretic mobility formed from the A and B subunits specified by the *LDH-A* and *LDH-B* loci (Fig. 1).

The breeding results for the parental generation in which the trisomic male was found (family M290), for two families produced by this male (0-37 and 0-38), and for families produced by two heterozygous male sibs (0-39 and 0-40) are presented in Table 1. The trisomic male arose in a family in which the proportions of offspring indicate the expected segregation of *LDH-B* alleles, if the trisomic offspring is excluded.

The trisomic male was testcrossed to two females of *BB* genotype to give the families 0-37 and 0-38. In each family the ratio of six offspring genotypes fit that expected if there had been random assortment of three chromosomes into equal numbers of functional *n* and *n* + 1 gametes. Moreover, the data from the two families are homogeneous enough to be combined ($\chi^2 = 0.82$, $P > .95$); as expected, the number of progeny with the six genotypes are nearly equal, as are the number of trisomic and disomic offspring (55 and 53, respectively). Typical zymogram patterns used to classify the six genotypes are shown in Fig. 1. The high concentration of the B_4 homotetramer in genotypes BBB'' (slot 1) and BBB' (slots 3 and 5) contrasts to the lower concentration of this tetramer in genotypes BB'' (slot 6) and BB' (slot 4). More bands are present in the zymogram from a $BB'B''$ fish (slot 2), and the second band from the origin in this zymogram is a doublet representing A_3B' and A_3B'' heterotetramers.

Two $B'B''$ male sibs from family M290 were testcrossed to BB females to produce the families 0-39 and 0-40 (Table 1); the ratios of offspring in both families are close to the 1 : 1 expected for offspring of disomic heterozygotes. Also, one female and three male sibs of the $B'B''$ genotype from family M290 were used in other crosses (1); all four produced results expected of normal disomic segregation. Thus, no unusual genetic results were recorded from six normal siblings of the trisomic male studied.

Cytological studies were made on the trisomic male after the breeding experiments and on some of his progeny from family 0-37. Squash preparations were made from gill and kidney tissues that were finely minced, treated with hypotonic saline, and fixed in Carnoy's fixative; the technique has been described (3). Chromosomes were stained with 1 percent aqueous crystal violet and photographed with a Leitz Ortholux photomicroscope. The normal brook trout modal chromosome number is $2n = 84$, with 100 total arms distributed as 16 metacentrics and 68 acrocentrics (3, 4). In contrast, all 25 cells examined from the trisomic male had two extra arms. The modal count (15 cells) was 85 chromosomes with 17 metacentrics and 68 acrocentrics. One cell had 87 chromosomes with 15 metacentrics and 72 acrocentrics; six cells had 86 chromosomes with 16 metacentrics and 70 acrocentrics; and three had 84 chromosomes with 18 metacentrics and 66 acrocentrics. While some of this variation might be due to counts made on disrupted cells from squash preparations, such intraindividual variation due to Robertsonian variation or centric fusion and fission is common in Salmonidae (3, 5). A karyotype of a cell showing the modal chromosome number with the extra metacentric appears in Fig. 2.

Among young fry from family 0-37,

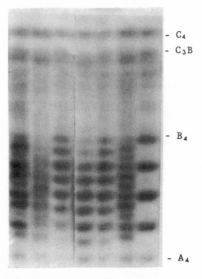

- C_4
- C_3B
- B_4
- A_4

B	B	B		B		
B	B'	B	B	B	B	B
B"	B"	B'	B'	B'	B"	B

Fig. 1. Starch gel zymograms of eye tissue LDH from progeny of a BB disomic female crossed with a $BB'B''$ trisomic male brook trout. Genotypes are designated at the bottom of each slot; homotetramers of A and B subunits of the ubiquitous system as well as tetramers of the C subunit of the eye system are shown along the side (origin and cathode at bottom; anode at top).

limited chromosome counts on known LDH genotypes were informative but not definitive. One metaphase in a *BB'B''* individual showed 102 arms distributed as 85 chromosomes with 17 metacentrics, a distribution identical to that of the male parent. Three *BBB'* individuals had a modal count of 85 chromosomes with 16 metacentrics, a total of 101 arms. Two cytological classes (with 100 and 101 arms, respectively) appeared among *BB''* disomics. All had modal counts of 84 chromosomes, but one group had 16 metacentrics and the other had 17 metacentrics. All other disomics had 100 arms with a modal count of 84 chromosomes (16 metacentrics and 68 acrocentrics).

The cytological and zymogram analyses show that the extra metacentric possessed by the original trisomic of genotype *BB'B''* was not an isochromosome. An isochromosome would have resulted in only one extra arm, and the ratios of genotypes and of disomics to trisomics among the progeny would have departed significantly from equality. If the extra metacentric were a trisomic isochromosome, the original male with three different alleles would have been tetrasomic for the chromosome bearing the *LDH-B* locus. Rather, the data indicate that the metacentric arose from centric fusion (Robertsonian translocation) of two acrocentric chromosomes, one of which carried the *LDH-B* locus. It is likely that the other acrocentric involved in the centric fusion was the chromosome carrying the *LDH-A* locus, because pseudolinkage of *LDH-A* and *LDH-B* alleles occurs among males of trout species and their hybrids and is correlated with centric fusion and fission (2, 3). If this hypothesis is correct, the trisomic chromosome set also possessed three doses of the *LDH-A* gene, because the normal complement of chromosomes was present in addition to the extra metacentric. In some zymograms the A$_4$ band appears heavier in trisomics than in disomics. However, the

lack of allelic differences for the *LDH-A* locus in the *BB'B''* trisomic made definitive analysis impossible.

The extra metacentric probably carried the *B''* allele. The one chromosome count made on a *BB'B''* offspring from family 0-37 showed 102 arms, including an extra metacentric; two *BB''* disomic offspring had an extra metacentric, while the extra chromosome in three *BBB'* progeny was an acrocentric. Also, all *BB'* offspring had 100 arms, whereas *BB''* offspring had cells with 101 and 102 arms as well. If *B''* was carried on the extra metacentric, the *BB''* disomics with the normal 84 chromosomes and no extra chromosome arms must have resulted from fission of the metacentric, followed by segregation of the *B''* acrocentric to one gamete and the *B* and *B'* acrocentrics to the other. The latter would give rise to *BBB'* offspring with one extra arm; three such offspring were found. The failure to find disomics for

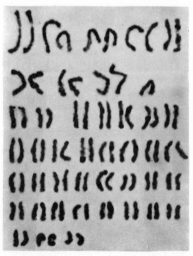

Fig. 2. Karyotype of a centric-fusion trisomic male brook trout. Two extra arms as a metacentric are at the end of the second line ($2n = 85$ with 17 metacentrics; fundamental arm number = 102). A satellite pair of chromosomes is in the third line.

Table 1. Breeding results from matings of brook trout of various *LDH-B* locus genotypes.

Family	Parental genotypes		Number of offspring with genotype:							P of χ^2
	♀	♂	BB	BB'	BB"	B'B"	BBB'	BBB"	BB'B"	
M290	BB"	BB'	7	9	16	12			1*	> .20
0-37	BB	BB'B"*	6	10	11		13	5	11	< .50
0-38	BB	BB'B"*	7	7	12		12	5	9	< .50
0-39	BB	B'B"		41	36					> .50
0-40	BB	B'B"		121	113					> .50

* Original trisomic.

LDH-B with an extra acrocentric (presumably bearing the *LDH-A* locus) may represent some univalent loss during parental gamete formation or may be simply the result of sample size. Unfortunately, we did not obtain data, such as cytological analysis of the *BBB"* progeny, that would have shown whether the metacentric carried the *B"* allele.

The equal numbers of the six genotypes and of disomics and trisomics among the progeny of the male trisomic suggest that meiosis in this male must have been precise. The metacentric bearing the *B"* allele probably formed a bivalent and a univalent randomly. Disjunction of the bivalent members plus random segregation and regular inclusion of the univalent in meiotic products must then have occurred. It is possible that other gametic combinations are produced but are lethal.

We had hoped to identify a specific chromosome as the bearer of the *LDH-B* locus and its linkage group. This was not possible because of the relative lack of differences among the 68 acrocentric chromosomes in a normal array (Fig. 2). New staining techniques may make such identification possible. We showed that aneuploidy can be superimposed on the Robertsonian variation extant in species of

Salmonidae and that it can be recognized when it involves chromosomes carrying alleles for biochemical traits with distinguishable genotypes.

Two additional points should be made. In most other animals, including humans, even partial trisomy is usually deleterious (6). In contrast, neither differences in size and viability nor abnormalities in phenotype or meiotic segregation were associated with trisomy in these fish. Second, it has been proposed that salmonids are ancient tetraploids. Recurrent trisomy may have been one mechanism for gradually attaining polyploidy and for creating a duplicated genome on which the selective processes of evolution could act (7). Many lines of evidence point to the importance of the flexible genomes of these fishes in evolution and in our understanding of evolutionary processes.

References and Notes

1. J. E. Wright and L. M. Atherton, *Genetics* **60**, 240 (1968).
2. W. J. Morrison and J. E. Wright, *J. Exp. Zool.* **163**, 259 (1966); W. J. Morrison, *Trans. Amer. Fish. Soc.* **99**, 193 (1970); J. E. Wright and L. M. Atherton, *ibid.*, p. 179.
3. M. T. Davisson, thesis, Pennsylvania State University (1969); ———, J. E. Wright, L. M. Atherton, in preparation.
4. G. Svärdson, *Report 23, Swedish State Institute of Freshwater Fisheries Research* (1945); J. E. Wright, *Progr. Fish. Cult.* **17**, 172 (1955).
5. S. Ohno, C. Stenius, E. Faisst, M. T. Zenges, *Cytogenetics* **4**, 117 (1965); S. Ohno, J. Mura-

moto, J. Klein, N. B. Atkin, *Chromosomes Today* **2**, 139 (1969); F. L. Roberts, *Trans. Amer. Fish. Soc.* **99**, 105 (1970); J. R. Heckman, F. W. Allendorf, J. E. Wright, *Science* **173**, 246 (1971).
6. P. E. Polani, *Brit. Med. Bull.* **25**, 81 (1969).
7. S. Ohno, *Evolution by Gene Duplication* (Springer-Verlag, New York, 1970), pp. 104, 107.
8. This is paper 4203 in the journal series of the Pennsylvania Agricultural Experiment Station in cooperation with the Benner Spring Fish Research Station, Pennsylvania Fish Commission. Supported by NSF grant GB4624 and by NIH predoctoral fellowship 4-Fol-GM-38853.

Extensive Heterozygosity at Three Enzyme Loci in Hybrid Sunfish Populations

Gregory S. Whitt, William F. Childers, John Tranquilli, and Michael Champion

Hybrid populations of sunfishes were produced in two different ponds, and the frequencies of allelic isozyme phenotypes were determined for three enzyme systems—malate dehydrogenase (NAD), esterases, and tetrazolium oxidase—in order to estimate the extent of heterozygosity at four different genetic loci. Interspecific F_1 hybrid fry (red-ear male × bluegill female) were produced in vitro. These fry were stocked in ponds at the free-swimming stage. When 1 year old, the F_1 hybrids produced a large F_2 hybrid population. Successful hybrid reproduction occurred each year thereafter. In one pond, a 1-year-old F_2 population exhibited all three isozyme phenotypes (red-ear, F_1, bluegill) at most loci in the approximate ratio of the $1:2:1$ expected. In a second pond, 5-year-old individuals of the F_2 generation were morphologically like the F_1 and were all heterozygous for the enzyme loci studied. This unusual degree of heterozygosity in the older F_2 population appeared to be the result of differential survival of mature heterozygous individuals and not the result of early embryonic lethality. The increased heterozygosity at these unlinked loci was assumed to reflect the condition at other genetic loci in the F_2 hybrids. Several possible mechanisms are advanced to explain this apparent heterosis.

INTRODUCTION

Some interspecific hybrids grow more rapidly and are more vigorous than either parent—a phenomenon termed "heterosis" (Shull, 1948). Heterosis is

This research was supported by NSF grant GB-16425 (G.S.W.) and by funds from the Illinois Natural History Survey (W.F.C.).

110

commonly attributed to the physiological superiority of heterozygotes. If, in fact, heterozygotes are superior to homozygotes, one might expect to observe an excess of heterozygotes in segregating hybrid populations. The use of isozymes as gene markers permits this question to be approached experimentally.

Heterosis at single enzyme loci has been demonstrated in various ways for both plants (Schwartz, 1960; Schwartz and Laughner, 1969) and animals (Manwell et al., 1963; Koehn, 1969, 1970; Manwell and Baker, 1969; Richmond and Powell, 1970). Some analyses of natural fish populations have also suggested heterosis. For example, an excess of heterozygotes was observed at the sorbitol dehydrogenase locus in a wild population of goldfish (Lin et al., 1969).

Interspecific hybridization of sunfishes (Centrarchidae) (Childers, 1967) has permitted an analysis of the molecular basis of heterosis (Manwell et al., 1963; Manwell and Baker, 1969, 1970). More recently, analyses of the inheritance of allelic isozymes in hybrid sunfishes have indicated that an excess of heterozygotes may exist for some loci (Wheat et al., 1971; Metcalf et al., 1972a) but not for others (Whitt et al., 1971; Wheat et al., 1972).

Interspecific sunfish hybrids are particularly useful in the study of heterosis because homologous loci of these species have diverged during evolution, and parental enzymes frequently have different electrophoretic mobilities as well as other physical and kinetic differences (Metcalf et al., 1972a,b; Whitt et al., 1971, 1972; Wheat et al., 1971, 1972). In addition, because the amount of genetic diversity is tremendously increased in interspecific hybrid populations, we might expect selection to proceed at a much faster rate than the rates of natural selection for polymorphisms within a given single species or population.

The purpose of the present investigation was to employ allelic isozymes to determine whether heterozygotes at several genetic loci are more frequent than could be expected by Hardy–Weinberg calculations and then to determine whether any deviations from the expected are consequences of early differential mortality or whether they result from selection acting upon mature individuals.

MATERIALS AND METHODS
Production of F_1 Interspecific Hybrids

Samples of eggs from several ripe female bluegills (*Lepomis macrochirus*) were stripped into petri dishes containing 20 ml of aged aerated tap water. The eggs were dispersed by gently shaking the dishes. Two drops of sperm from a ripe male red-ear sunfish (*L. microlophus*) were stripped into each petri dish. After the zygotes had been washed free of excess sperm, they were maintained in constantly aerated water at approximately 30 C. Dead embryos were removed as soon as they were detected.

Pond Stocking of the F_1 Fry

Six days after fertilization, the newly hatched F_1 larvae had developed into free-swimming fry. In 1964, approximately 22,000 F_1 hybrid fry were stocked in Bluebird Pond, located in east-central Illinois. At the time of stocking, no other species of fish were in the pond. However, 2 years after stocking the F_1 hybrids, small numbers of largemouth bass and F_1 hybrid crappie were also stocked in the pond. Although both of these species probably exerted some predation on the red-ear × bluegill hybrids, they did not interbreed with them. Approximately 99% of these F_1 hybrids were males, which according to Haldane's rule (Haldane, 1922) suggests that the female is the heterogametic sex. The F_1 hybrids produced F_2 and successive (F_2+) generations. In the fall of 1970, a sample of 380 fish was removed from this pond by electronarcosis. This sample was estimated to represent 1–5% of all fishes in the pond.

Employing essentially identical procedures as described for Bluebird Pond, approximately 5000 F_1 hybrid fry were stocked in Sullivan Lab Pond in 1969. This pond also contained channel catfish and largemouth × smallmouth bass hybrids, which probably exerted predation pressure on the red-ear × bluegill hybrids. In the fall of 1971, a sample of 103 F_2 and F_2+ individuals was collected from this pond.

Morphological Determinations

The year class (and generation) of each individual fish was determined by counting the number of annular growth rings (annuli) on scales removed from the middle of the left side below the lateral line. The standard length and sex of each fish were determined.

A morphological index was used to evaluate three morphological characteristics. A value of 6 was assigned to the pure bluegill phenotype and 10 to the pure red-ear phenotype. Intermediate phenotypes were assigned intermediate values. The three morphological characteristics scored were the color of the opercular tab (on the gill flap), the color on the side of the head, and the morphology of the gill rakers.

Homogenization Procedure

Caudal skeletal muscle of each frozen fish was dissected and then hand-homogenized in 1 vol of tris–HCl buffer (0.1 M, pH 7.0) at 4 C. The homogenate was centrifuged for 30 min in a Sorvall Superspeed RC2-B centrifuge at 4 C and $48,200 \times g$. The supernatant was recentrifuged under the same conditions, and the final supernatant fraction was immediately subjected to electrophoresis.

Electrophoresis

Vertical starch gel electrophoresis (Buchler Instruments) was done in 14% gels (electrostarch) at 8 v/cm for 16 hr. A tris–citrate pH 6.8 stock buffer of 0.75 M tris and 0.25 M citrate (monohydrate) was diluted 1:30 for gels and 1:7 for both electrode chambers.

The staining procedures were those of Shaw and Prasad (1970). Tetrazolium oxidase was detected by the addition of 10 ml PMS (1.6 mg/ml) and 10 ml NBT (1 mg/ml) to 100 ml 0.5 M pH 9.0 tris–HCl.

RESULTS

Morphology of the Red-Ear, Bluegill, and F_1 Hybrid Sunfishes

The two parental species and the F_1 interspecific hybrid are shown in Fig. 1. The bluegill possesses a dark bluish-black opercular tab and a dark spot on the soft dorsal fin. The red-ear sunfish lacks the spot on the dorsal fin and possesses a scarlet border on the opercular tab. The morphology of the F_1 interspecific hybrid is generally intermediate to those of the two parental species.

Morphology of the F_2 and F_{2^+} Hybrids

The sex and morphological indices for three characteristics of 380 individuals from Bluebird Pond are given in Table I. All year classes were present in the sample, but some year classes (1968, 1970) were present in very low numbers.

All the 6-year-old fish (as determined by the number of annuli) were those F_1 individuals originally introduced into the unpopulated pond. All the 1965 year class fish (as determined by the number of annuli) had to be F_2 individuals because the only fish in the pond capable of being their parents were F_1 hybrids. On the basis of their age, morphology, and isozyme composition the 1966 generation also appeared to be F_2 hybrids. The F_2 hybrids were well represented, with a total of 236 individuals of a sample of 380. The year classes 1967, 1968, 1969, and 1970 are referred to as F_{2^+} because they could have been formed by these following crosses: $F_1 \times F_1$, $F_1 \times F_2$, $F_2 \times F_2$, and $F_{2^+} \times F_{2^+}$.

Our samples of the F_1 and F_2 generations contained 100% males. However, estimates based on much larger samples of the F_1 individuals of this

113

Fig. 1. F_1 interspecific hybrid (middle), bluegill (top), and red-ear sunfish (bottom).

cross indicated that males constitute only 99.4% of the total population. The presence of 0.6% females was sufficient to produce a sizeable F_2 population. A striking change in the sex ratio occurred in the F_{2+} generations, which revealed approximately 50% males in each year class.

As expected, the F_1 interspecific hybrids possessed morphological phenotypes intermediate to those of the two parental species. The morphologies of the F_2 individuals were diverse, representing the complete spectrum from one parental type to the other. However, a majority of the F_2 generation exhibited phenotypes similar to those of the F_1 (i.e., intermediate).

A striking change in morphology occurred in the 1967 year class and persisted for all of the F_{2+} generation fish. All of the F_{2+} generation individuals possessed pure bluegill morphological phenotypes. Thus the F_1 and F_2 generations possessed very similar morphologies and sex ratios, whereas the F_{2+} generations differed from these generations in both aspects.

In addition to the abrupt transition from the F_2 to the F_{2+} generations in morphology and sex ratios, there was also a marked change in the ratio of allelic isozyme phenotypes.

114

Table I. Sex and Morphology of Individuals from the Male Red-Ear × Female Bluegill Hybrid Population from Bluebird Pond

Year class[a]	Age of the fish (years)	Generation[b]	No.	Percent males in sample[c]	Standard length (mm)		Morphological index[d]					
							Color of opercular tab		Color on side of head		Gill raker morphology	
					$\bar{X}$	SD	$\bar{X}$	SD	$\bar{X}$	SD	$\bar{X}$	SD
1964	6	F_1	62	100.0	113.4	20.3	7.37	0.78	7.66	0.83	8.35	0.60
1965	5	F_2	216	100.0	99.1	6.5	7.56	0.76	7.67	0.76	8.16	0.54
1966	4	F_2	20	100.0	92.8	4.1	7.30	0.80	7.35	0.81	7.90	0.91
1967	3	F_2+	48	45.8	80.3	5.5	6.0	0	6.0	0	6.0	0
1968	2	F_2+	9	44.4	73.4	2.6	6.0	0	6.0	0	6.0	0
1969	1	F_2+	24	45.8	71.8	4.7	6.0	0	6.0	0	6.0	0
1970	—	F_2+	1	—	28.0	—	—	—	—	—	—	—

Total = 380

[a] Year class based on a count of the number of annuli on scales.

[b] F_2+ indicates fish could result from $F_1 \times F_1$, $F_1 \times F_2$, $F_2 \times F_2$, etc.

[c] Although sample indicates 100% males for F_1, estimates based on much larger sample sizes indicate there are 99.4% males in total population.

[d] Bluegill phenotype, 6; red-ear phenotype, 10.

115

Isozyme Phenotypes of the Bluegill, Red-Ear and F_1 Hybrids

Enzyme phenotypes resolved by starch gel electrophoresis are shown in Figs. 2, 3, and 4 for malate dehydrogenase, tetrazolium oxidase, and esterases. All of the enzyme extracts were from skeletal muscle.

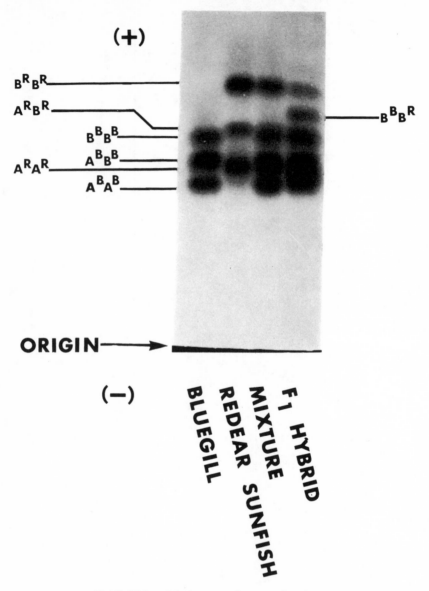

Fig. 2. Malate dehydrogenase isozyme phenotypes.

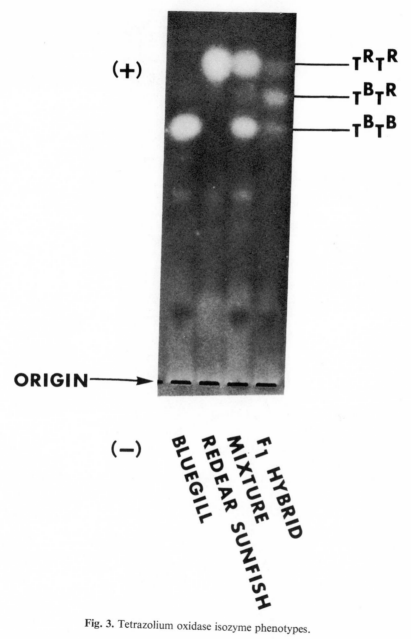

Fig. 3. Tetrazolium oxidase isozyme phenotypes.

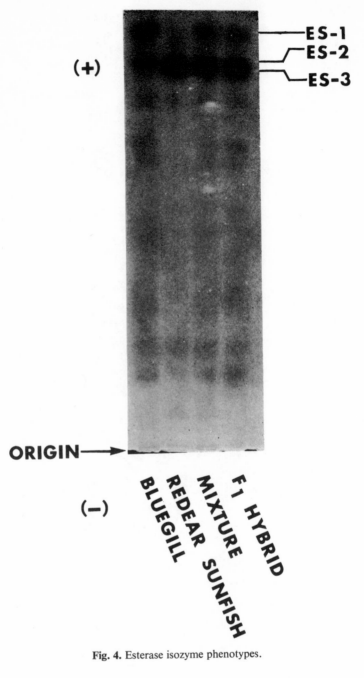

Fig. 4. Esterase isozyme phenotypes.

The supernatant malate dehydrogenase (MDH) (E.C. 1.1.1.37) was composed of a set of three isozymes in each species (Fig. 2). Each species had two homopolymers, AA and BB, and one heteropolymer, AB, of intermediate electrophoretic mobility. The mobilities of parent-species isozymes were different. The F_1 hybrid possessed all the isozymes of both parental species, including unique heterodimers composed of subunits contributed by each species, e.g., $B^R B^B$ (see Wheat et al., 1971, 1972, and Wheat and Whitt, 1971, for the molecular and genetic bases of sunfish MDHs).

The tetrazolium oxidase or "diformazan oxidase" phenotypes are shown in Fig. 3. The skeletal muscle of each parental species had a single main band of activity. The mixture of the two parental-species enzyme extracts resulted in the two parental enzymes after electrophoresis. However, in the F_1 hybrids, the two parental forms were present as well as a band of intermediate electrophoretic mobility.

Esterases of the two species (Fig. 4) also possessed different electrophoretic mobilities. The bluegill had an anodal esterase (ES-1) which the red-ear lacked, while the red-ear sunfish possessed a cathodal isozyme (ES-3) which was absent in the bluegill. A mixture of the two parental extracts resulted in an isozyme pattern essentially identical to that of the F_1 hybrid.

Frequency of Allelic Isozyme Phenotypes in the Successive Hybrid Generations

Because the allelic isozymes are encoded in codominant alleles, we were readily able to determine genotypes at all three enzyme loci for individuals from the Bluebird Pond population (Table II). The MDH-A phenotypes from this population were not sufficiently resolved by the electrophoresis to be analyzed. As expected, all F_1 individuals (1964 year class) possessed the F_1 hybrid phenotype for each enzyme.

Unexpectedly, however, all individuals of the F_2 generations were also heterozygous at all three enzyme loci. All of the 216 F_2 individuals (1965 year class) possessed only the F_1 hybrid phenotype instead of the expected $1:2:1$ ratio of phenotypes, specifically 54 red-ear : 108 F_1 hybrid : 54 bluegill. The difference between the observed ratios and those expected was highly significant.

It was only in the F_{2+} generations that individuals homozygous for an enzyme phenotype were observed. However, of the two parental phenotypes possible, only those of the bluegill were detected. The ratio of the distribution of isozyme phenotypes observed is significantly different ($P < 0.01$) from that expected (MDH-B $\chi^2 = 20.5$, 2 d.f.; esterase $\chi^2 = 22.9$, 2 d.f.; tetrazolium oxidase $\chi^2 = 16.5$, 2 d.f.). The phenotypic composition of the 1968 year class was not significantly different ($P > 0.05$) from that expected (MDH-B $\chi^2 = $

119

Table II. Distribution of Isozyme Phenotypes in the Male Red-Ear×Female Bluegill Hybrid Population from Bluebird Pond

Class	Age	Generation	No.	Percent MDH B phenotypes			Percent esterase phenotypes			Percent tetrazolium oxidase phenotypes		
				Red-ear	F_1 hybrid	Bluegill	Red-ear	F_1 hybrid	Bluegill	Red-ear	F_1 hybrid	Bluegill
1964	6+	F_1	62	—	100	—	—	100	—	—	100	—
1965	5+	F_2	216	—	100	—	—	100	—	—	100	—
1966	4+	F_2	20	—	100	—	—	100	—	—	100	—
1967	3+	F_2+	48	—	54	46	—	52	48	—	71	29
1968	2+	F_2+	9	—	67	33	—	44	56	—	56	44
1969	1+	F_2+	24	—	71	29	—	46	54	—	63	37
1970	0+	F_2+	1	—	100	—	—	—	100	—	—	100

Table III. Distribution of Isozyme Phenotypes in the Male Red-Ear × Female Bluegill Hybrid Population from Sullivan Lab Pond

Age (years)	Generation	No.	Percent MDH A phenotypes			Percent MDH B phenotypes			Percent esterase phenotypes			Percent tetrazolium oxidase phenotypes		
			Red-ear	F_1 hybrid	Bluegill	Red-ear-	F_1 hybrid	Bluegill	Red-ear	F_1 hybrid	Bluegill	Red-ear	F_1 hybrid	Bluegill
1+	F_2	35	11	57	32	29	46	25	9	57	34	—	60	40[a]
0+	F_2+	68	4	48	48[a]	19	72	9[a]	—	53	47[a]	—	55	45[a]

[a] Significantly different ($P < 0.01$) from the expected 1:2:1 ratio.

121

3.0, 2 d.f.; esterase $\chi^2 = 4.6$, 2 d.f.; tetrazolium oxidase $\chi^2 = 3.7$, 2 d.f.). The composition of the 1969 year class isozymes was significantly different from that expected ($P<0.05$) (MDH-B $\chi^2 = 8.3$, 2 d.f.; esterase $\chi^2 = 14.3$, 2 d.f.; tetrazolium oxidase $\chi^2 = 8.3$, 2 d.f.).

Contingency χ^2 tests indicate that the isozyme alleles at each locus in the F_{2^+} generations assort randomly with respect to sex and each other, indicating that these genetic loci are probably not linked to each other or to sex. Identical results were also observed for the second hybrid population studied.

The great excess of heterozygotes observed in the F_2 generation suggested a differential survival of individuals heterozygous at these enzyme loci. The selection could, however, have occurred at any stage from fertilization to maturity. To explore this issue, we examined a younger F_2 hybrid population produced in a different pond (Sullivan Lab Pond).

Frequency of Allelic Isozyme Phenotypes in the Young F_2 and F_{2^+} Hybrids

The second phase of the study was initiated in 1969 when 5000 F_1 fish were placed in Sullivan Lab Pond to give rise to an F_2 generation in 1970 and an F_{2^+} generation in the spring of 1971. This pond was sampled in the fall of 1971, and the isozyme phenotypes of 103 fish were determined.

In the F_2 generation of this sample, all three isozyme phenotypes (red-ear, F_1 hybrid, bluegill) were present for each of three enzyme systems (MDH-A, MDH, B, and esterase) (Table III), whereas only the F_1 and bluegill phenotypes were observed for tetrazolium oxidase. At the first three loci, observed phenotypic ratios did not deviate significantly ($P>0.05$) from the expected ratio (MDH-A $\chi^2 = 3.5$, 2 d.f.; MDH-B $\chi^2 = 0.4$, 2 d.f.; esterase $\chi^2 = 5.4$, 2 d.f.). However, the ratio for tetrazolium oxidase was significantly different ($P<0.01$) with $\chi^2 = 12.6$, 2 d.f.

In the F_{2^+} generation, the observed phenotypic distribution was significantly different from a $1:2:1$ ratio ($P<0.01$) for all four enzyme systems (MDH-A $\chi^2 = 28.3$, 2 d.f.; MDH-B $\chi^2 = 14.7$, 2 d.f.; esterase $\chi^2 = 30.4$, 2 d.f.; tetrazolium oxidase $\chi^2 = 28.8$, 2 d.f.). Although homozygous individuals were present in the young F_2 and F_{2^+} generations from Sullivan Lab Pond, the frequencies of red-ear sunfish isozyme phenotypes were lower than expected for most of the enzymes.

DISCUSSION

Isozyme Phenotypes

Malate dehydrogenase isozymes of the red-ear and bluegill are similar to the three supernatant MDH isozymes reported for other sunfish (Whitt *et al.*,

1971, 1972; Wheat et al., 1971, 1972; Wheat and Whitt, 1971) and appear to be dimers resulting from the random assembly of two different (A and B) polypeptides (Wheat et al., 1971). The MDH-A and MDH-B subunits are encoded in genetic loci which are unlinked (Wheat et al., 1972).

Tetrazolium oxidase phenotypes are like those reported for other sunfish (Whitt et al., 1972). The possession of three bands in the F_1 hybrid suggests a dimeric enzyme.

Skeletal muscle esterases are similar in their electrophoretic mobility and staining intensity to those observed in the white crappie and black crappie (Metcalf et al., 1972a,b). Subunit composition of these isozymes was not determined.

In the F_1 hybrids, both parental alleles were functioning at each enzyme locus examined, unlike the allelic repression observed in the F_1 interspecific sunfish hybrid (red-ear × warmouth) (Whitt et al., 1972).

The random assortment of the bluegill and red-ear alleles at each locus in the F_2^+ generations in both hybrid populations indicates that these enzyme loci are not linked.

Distribution of Isozyme Phenotypes in the Hybrid Population

The F_1 and F_2 generations in the older hybrid population (Bluebird Pond) both exhibited a skewed sex ratio, intermediate morphology, and 100% heterozygosity for all enzyme loci examined. Abruptly at the F_2^+ generation there were several striking changes in these attributes. The F_2^+ population possessed a normal sex ratio, pure bluegill morphology, and a decrease in heterozygosity for all of the enzyme systems.

This tendency toward homozygosity in the F_2^+ generation might be explained in several ways. It may represent a "hybrid breakdown" in the F_2^+ generation due, perhaps, to separation of coadapted gene complexes in the meiosis of the F_2. Another possibility is that only one F_2 female, homozygous for bluegill isozymes and morphology, participated in a mating with either an F_1 male or an F_2 male heterozygous for the three enzyme systems. This latter situation might explain the exclusive presence of F_1 hybrid and bluegill isozyme phenotypes in the F_2^+ generations.

The data from the older F_2 generation from Bluebird Pond are certainly suggestive of heterozygote selection, if we can assume that the unlinked enzyme loci studied are representative of other loci throughout the genome.

F_2 individuals from Bluebird Pond were approximately 5 years old when examined. Thus it was not known whether the selection for complete heterozygosity occurred early in life (embryonic inviability) or later in the life cycle. In addition, the numbers of F_2^+ individuals in Bluebird Pond were very small, especially for the younger year classes, e.g., only one individual in the

1970 year class. Alternatively, 103 fish were obtained from Sullivan Lab Pond (35 F_2, 68 F_2+), and the data from this sample do permit some strong conclusions. The presence of segregant isozyme phenotypes in the F_2 generation in the expected ratio of $1:2:1$ certainly suggests that the high degree of heterozygosity observed in older F_2 individuals from Bluebird Pond is the result of natural selection acting over a period of several years.

In both the F_2 and F_2+ generations of the younger hybrid population, the frequency of red-ear phenotypes is lower than expected. This observation is consistent with the data from the F_2+ generation of the older hybrid population, which completely lacked red-ear phenotypes.

A surprising observation was that in the younger hybrid population from Sullivan Lab Pond the F_2+ individuals (1971 year class) deviated more dramatically from the expected ratio than did the older F_2 individuals. This suggests that in the F_2+ generation some selection occurs early in the life cycle as well as over a long term. In a few more years we will be able to determine if F_2+ individuals 5–6 years of age are heterozygous for all three enzyme systems— as were the F_2 individuals of the same age.

If both the old and young hybrid populations are considered together, the following rationale for the observed change in allelic isozyme frequencies can be postulated. The very young F_2 fry probably possess a $1:2:1$ phenotypic ratio at each locus, produced by normal segregation. As the F_2 individuals age during the first few years, the homozygous red-ear isozyme phenotypes are selected against. In later years, the homozygous bluegill isozyme phenotypes are selected against, leaving only heterozygous individuals in the older F_2 population. We realize the large magnitude of genetic death that must occur by this scheme. It must be remembered that we are dealing with a genetically unique situation and that these magnitudes of selection surely do not exist in genetically variable, but nonhybrid, natural populations.

We have assumed that the behavior of the three-enzyme system's phenotypes in the hybrid population is representative of allelic isozymes at many loci. We do not yet know what mechanisms are involved in the selection of F_2 individuals with an F_1 hybrid phenotype. However, we have postulated some alternative hypotheses to explain the apparent enhanced survival of the more heterozygous F_2 individuals.

Although it is known that the F_1 sunfish hybrids grow more rapidly and are more aggressive than the individuals of the parental species, it is not clear whether these attributes are related to the apparent advantage shown by the heterozygous F_2 fish that are more genetically similar to F_1s. It is conceivable that the F_2 fish with an F_1 phenotype are more efficient in competition for food. Alternatively, the more heterozygous F_2 individuals may be more resistant to disease and predation than the more homozygous F_2 fish.

A hypothesis that could explain the selection for heterozygous individuals in the F_2 population is that natural selection favors balanced chromosome sets like those of the F_1 hybrid and highly heterozygous F_2 individuals.

The properties of the allelic isozymes for these three enzymes are not necessarily responsible for the selection for or against an individual. It may be that the increased heterozygosity at these three loci in the F_2 generation simply reflects a selection for heterozygosity at many other genetic loci. One example of this would be selection for balanced chromosome sets characteristic of the F_1.

Alternatively, the existence of heterozygosity at many loci may confer a physiological advantage to the F_1 hybrid or to F_2 individuals that are heterozygous for many enzyme loci.

One possible molecular mechanism to explain this heterozygous advantage has been advanced by Manwell et al. (1963), who observed that the hybrid hemoglobin molecules (containing subunits encoded in both species) of some F_1 hybrid sunfish possess a higher affinity for oxygen than the hemoglobin of either parental species. This "overdominant" effect of polypeptides in the hybrid polymer was postulated to be responsible for the hybrid vigor in the F_1. Perhaps similar subunit interactions occur for many polymeric enzymes in heterozygous F_1 and F_2 individuals. A similar model system has been studied within another species of fish (*Catostomus clarki*) by Koehn (1969, 1970). The allelic esterase activities had different temperature optima. In a cline, temperature appears to be the selective force determining the esterase allele frequencies. The higher degree of heterozygosity in individuals over part of the cline appears to be due to the physiological advantage of the heterozygote when it occupies an intermediate temperature environment. In the case of the sunfish, if the enzymes from the bluegill and red-ear are effective over different ranges of temperature (or other factors) because of differences in their environments, then perhaps the possession of both parental enzymes may permit heterozygous individuals to operate more efficiently over a broader range than either of the parental species.

Because of the considerable genetic divergence of the two species studied, we would expect considerable genetic diversity in the F_2 generation. Therefore, a much more stringent and rapid selection would be expected in these hybrid populations than that observed for enzyme polymorphisms within a single given species.

Future investigations will attempt to determine whether significant differences exist in the physical and kinetic properties of the different allelic forms of the sunfish enzymes. If differences occur, then the heterozygous individuals will be subjected to specific selection pressures to determine whether they possess a selective advantage.

ACKNOWLEDGMENTS

The authors thank G. W. Bennett and D. L. Nanney for their helpful suggestions. We are grateful to R. K. Koehn for his critical evaluation of the manuscript. The photographs were taken by W. D. Zehr of the Illinois Natural History Survey.

REFERENCES

Childers, W. F. (1967). Hybridization of four species of sunfishes (Centrarchidae). *Ill. Nat. Hist. Surv. Bull.* **29**:159.

Haldane, J. B. S. (1922). Sex ratio and unisexual sterility in hybrid animals. *J. Genet.* **12**:101.

Koehn, R. K. (1969). Esterase heterogeneity: Dynamics of a polymorphism. *Science* **163**:943.

Koehn, R. K. (1970). Functional and evolutionary dynamics of polymorphic esterases in catostomid fishes. *Trans. Am. Fisheries Soc.* **99**:219.

Lin, C. C., Schipmann, G., Kittrell, W. A., and Ohno, S. (1969). The predominance of heterozygotes found in wild goldfish of Lake Erie at the gene locus for sorbitol dehydrogenase. *Biochem. Genet.* **3**:603.

Manwell, C., and Baker, C. M. A. (1969). Hybrid proteins, heterosis and the origin of species. I. Unusual variation of polychaete *Hyalinoecia* "nothing dehydrogenases" and of quail *Coturnix* erythrocyte enzymes. *Comp. Biochem. Physiol.* **28**:1007.

Manwell, C., and Baker, C. M. A. (1970). *Molecular Biology and the Origin of Species: Heterosis, Protein Polymorphism and Animal Breeding*, Sidgwick and Jackson, London, 394 pp.

Manwell, C., Baker, C. M. A., and Childers, W. (1963). The genetics of hemoglobin in hybrids. I. A molecular basis for hybrid vigor. *Comp. Biochem. Physiol.* **10**:103.

Metcalf, R. A., Whitt, G. S., and Childers, W. F. (1972a). Inheritance of esterases in the white crappie (*Pomoxis annularis* Rafinesque), black crappie (*P. nigromaculatus* Lesueur) and their F_1 and F_2 interspecific hybrids. *Anim. Blood Grps. Biochem. Genet.*, **3**:19.

Metcalf, R. A., Whitt, G. S., Childers, W. F., and Metcalf, R. L. (1972b). A comparative analysis of the tissue esterases of the white crappie (*Pomoxis annularis* Rafinesque) and black crappie (*Pomoxis nigromaculatus* Lesueur) by electrophoresis and selective inhibitors. *Comp. Biochem. Physiol.* **41** [1B]:27.

Richmond, R. C., and Powell, J. R. (1970). Evidence of heterosis associated with an enzyme locus in a natural population of *Drosophila*. *Proc. Natl. Acad. Sci.* **67**:1264.

Roberts, F. L. (1964). A chromosome study of twenty species of Centrarchidae. *J. Morphol.* **115**:401.

Schwartz, D. (1960). Genetic studies on mutant enzyme in maize: Synthesis of hybrid enzymes by heterozygotes. *Proc. Natl. Acad. Sci.* **46**:1210.

Schwartz, D., and Laughner, W. J. (1969). A molecular basis for heterosis. *Science* **166**:626.

Shaw, C. R., and Prasad, R. (1970). Starch gel electrophoresis of enzymes—a compilation of recipes. *Biochem. Genet.* **4**:297.

Shull, G. H. (1948). What is "Heterosis"? *Genetics* **33**:439.

Wheat, T. E., and Whitt, G. S. (1971). *In vivo* and *in vitro* molecular hybridization of malate dehydrogenase isozymes. *Experientia* **27**:647.

Wheat, T. E., Childers, W. F., Miller, E. T., and Whitt, G. S. (1971). Genetic and *in vitro* molecular hybridization of malate dehydrogenase isozymes in interspecific bass (*Micropterus*) hybrids. *Anim. Blood Grps. Biochem. Genet.* **2**:3.

Wheat, T. E., Whitt, G. S., and Childers, W. F. (1972). Linkage relationships between the homologous malate dehydrogenase loci in teleosts. *Genetics* **70**:337.

Whitt, G. S., Childers, W. F., and Wheat, T. E. (1971). The inheritance of tissue speci
lactate dehydrogenase isozymes in interspecific bass (*Micropterus*) hybrids. *Bioche
Genet.* **5**:257.

Whitt, G. S., Cho, P. L., and Childers, W. F. (1972). Preferential inhibition of alle
isozyme synthesis in an interspecific sunfish hybrid. *J. Exptl. Zool.* **179**:271.

Phosphoglucose Isomerase Gene Duplication in the Bony Fishes: An Evolutionary History

John C. Avise   and G. Barrie Kitto

Electrophoretic patterns of phosphoglucose isomerase (PGI) in bony fishes provide strong evidence for a model of genetic control by two independent structural gene loci, most likely resulting from a gene duplication. This model is confirmed by a comparison of certain kinetic and molecular properties of the PGI homodimers (PGI-1 and PGI-2) isolated from extracts of the teleost Astyanax mexicanus. *In addition, in most higher teleosts examined, the PGI enzymes show a regular pattern of tissue distribution, with PGI-2 predominant in muscle, the heterodimer often strongest in the heart, and PGI-1 predominant in liver and other organs. An examination of 53 species of bony fishes belonging to 38 families indicates a widespread occurrence of duplicate PGI loci and an early origin of the gene duplication, perhaps in the Leptolepiformes. The apparent presence of three PGI loci in trout and goldfish exemplifies how new loci can be incorporated into the genome through polyploidization.*

INTRODUCTION

Almost all vertebrates and invertebrates thus far studied possess only a single phosphoglucose isomerase (PGI) gene locus: man (Detter *et al.*, 1968), deer (Ramsey *et al.*, 1972), rodents (Carter and Parr, 1967; Selander

This research was supported in part by a NSF graduate traineeship to J.C.A., by the Clayton Foundation for Research in Biochemistry (G.B.K.), by NSF Grant GB-15644 and NIH Grant GM-15769 to Robert K. Selander, and by contract AT(38-1)-310 between the University of Georgia and the U.S. Atomic Energy Commission.

et al., 1969; DeLorenzo and Ruddle, 1969; Johnson and Selander, 1971; Johnson *et al.*, 1972), birds (Nottebohm and Selander, 1972), lizards (Webster *et al.*, 1972), frogs (Ralin *et al.*, 1972), horseshoe crabs (Selander *et al.*, 1970), fiddler crabs (Selander *et al.*, 1972), and certain insects (Avise and Selander, 1972; Ramsey and Avise, unpublished).

Yoshida and Carter (1969) observed multiple electrophoretic bands of phosphoglucose isomerase, representing at least two structurally different isozymes, in rabbit hemolysates and muscle extracts. However, the number of genetic loci controlling PGI was not determined since all specimens showed identical patterns on the gels. Similarly, the genetic basis of three isozymes observed by Nakagawa and Noltmann (1967) in brewers and bakers yeast was not determined.

Avise and Selander (1972) describe allozymic patterns of phosphoglucose isomerase in a characid fish (*Astyanax mexicanus*) that are consistent with a model of control by two independent structural gene loci, most likely resulting from a gene duplication. In this paper, we present the evidence on which this conclusion is based, including an analysis of certain kinetic and molecular properties of the PGI homodimers in *A. mexicanus*. On the basis of a study of the tissue distribution and occurrence of multiple PGI loci in a wide variety of bony fishes (Osteichthyes), we propose a hypothesis of the time of origin of the gene duplication and the course of the divergent specialization of the homodimeric and heterodimeric enzyme forms.

MATERIALS AND METHODS

Specimens of *A. mexicanus* used in the extraction of PGI for characterization were collected in Cueva del Pachon, near Ciudad Mante, Tamaulipas, Mexico (see Avise and Selander, 1972). Other species were collected in Texas and Massachusetts or purchased from commercial dealers. All fish were stored on dry ice immediately after capture.

Fructose 6-phosphate (F6P), glucose 6-phosphate dehydrogenase (G6PDH), triphosphopyridine nucleotide (NADP), phenazine methosulfate (PMS), MTT tetrazolium (MTT), 6-phosphogluconic acid, and EDTA disodium salt were obtained from Sigma Chemical Company, phosphoenol pyruvic acid from Calbiochem, adenosine 5'-triphosphate from Pabst Laboratories Biochemical, and Electrostarch lot 171 from Otto Hiller, Madison, Wisconsin. Other chemicals were of reagent grade.

Starch Gel Electrophoresis

Extracts for electrophoresis were prepared by homogenizing the whole animal or tissue in a glass tissue grinder with an equivalent volume of buffer

(0.01 M tris, 0.01 M EDTA, 5×10^{-5} M NADP, pH adjusted to 6.8 with HCl). The homogenate was centrifuged at $49,000 \times g$ for 40 min, and the supernatant was stored at -60 C.

Horizontal starch gel electrophoresis was carried out as described by Selander *et al.* (1971, appendix), using either a Poulik (pH 8.7)–borate (pH 8.2) discontinuous buffer system at 25 v/cm for $2\frac{1}{2}$ hr, or a potassium phosphate (pH 6.7) continuous buffer at 13 v/cm for 6 hr. The enzyme was localized with a staining mixture described by Selander *et al.* (1971). All PGI bands appear anodal to the origin.

Isolation of the Homodimers

Isolation of the *Astyanax* PGI homodimers was carried out using an electrophoretic technique similar to that described by Prasad and Wright (1971). After electrophoresis, one slice of the gel was stained for PGI activity. The stained slice was then superimposed on the corresponding remainder of the gel, and a cut was made through the gel just anodal to each homodimer, using the stained slice as a guide. A small sponge was placed in the slit, and a strip of dialysis tubing was inserted between the sponge and the gel, just anodal to the sponge. The gel was then electrophoresed for $\frac{1}{2}$ hr, allowing each homodimer to run up into its appropriate sponge (the dialysis tubing prevented the enzyme from leaving the sponge). The enzyme was recovered by washing the sponge in a small volume of buffer.

Enzyme Assay

Enzymatic activity was measured at room temperature by the spectrophotometric assay procedure of Noltmann (1964). A Zeiss PMQ II spectrophotometer was used to follow the change in absorbance at 340 mμ caused by NADPH formation in a coupled enzyme system (F6P substrate and G6PDH indicator enzyme). The assay solution before addition of phosphoglucose isomerase enzyme solution contained 0.138 M tris, 0.023 M $MgCl_2$, 4×10^{-4} M F6P, 7.4×10^{-5} M NADP, and 1.96 units G6PDH. Addition of 0.1 ml PGI solution brought the final assay volume to 3 ml.

The activity of each of the two forms of PGI was measured over varying concentrations of inhibitor. The inhibitor and PGI were allowed to mix for 1 min before being added to the remainder of the assay solution. Michaelis constants for PGI homodimers were determined by varying the concentration of F6P.

Heat Inactivation

Thermal stabilities of the PGI homodimers were studied by heating the enzymes (a) at different temperatures for 5 min and (b) at 45 C for different

131

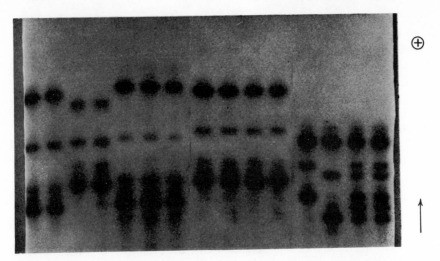

Fig. 1. Phenotypes of phosphoglucose isomerases from whole-animal extracts. Left to right: Two homozygous bluegill, two homozygous orangespotted sunfish, three homozygous silversides, four homozygous largemouth bass, and four mosquitofish with presumed phenotypes as follows: $Pgi\text{-}1^a/Pgi\text{-}1^a$, $Pgi\text{-}2^a/Pgi\text{-}2^a$; $Pgi\text{-}1^a/Pgi\text{-}1^a$, $Pgi\text{-}2^b/Pgi\text{-}2^b$; $Pgi\text{-}1^a/Pgi\text{-}1^a$, $Pgi\text{-}2^a/Pgi\text{-}2^b$; and $Pgi\text{-}1^a/Pgi\text{-}1^a$, $Pgi\text{-}2^a/Pgi\text{-}2^b$.

periods. After heating, the enzyme solutions were chilled on ice, and remaining PGI activity was assayed at room temperature.

RESULTS

Electrophoretic Analysis

In fish tissues, the patterns of PGI activity on the gels are similar to those for the multiple loci encoding malate dehydrogenase in several fishes (Bailey *et al.*, 1970; Whitt, 1970; Avise and Selander, 1972). Typically, individuals homozygous at both PGI loci show three widely spaced bands (see Fig. 1). (Other paler bands slightly anodal to the slowest are interpreted as secondary or satellite bands.) Presumably, the three bands result from association of two types of polypeptide subunits, one type encoded by each locus, to produce AA, AB, and BB dimers. The faster-migrating, more anodal band represents the homodimer produced by association of A subunits encoded by locus *Pgi-1*, and the slower band represents the homodimer produced by association of B subunits encoded by locus *Pgi-2*. (In the following discussion, PGI-1 and PGI-2 will specifically refer to the homodimeric PGI enzymes encoded by *Pgi-1* and *Pgi-2*, respectively.) Since the intermediate heterodimer band stains less intensely than those representing the homodimers, the association

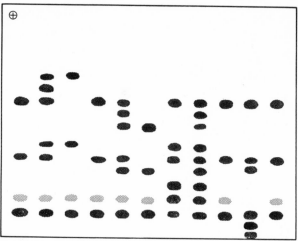

Fig. 2. Diagrammatic representation of some of the phenotypes observed in whole-animal extracts of *A. mexicanus*. Presumed phenotypes from left to right: $Pgi\text{-}1^b/Pgi\text{-}1^b$, $Pgi\text{-}2^b/Pgi\text{-}2b$; $Pgi\text{-}1^a/Pgi\text{-}1^b$, $Pgi\text{-}2^b/Pgi\text{-}2^b$; $Pgi\text{-}1^a/Pgi\text{-}1^a$, $Pgi\text{-}2^b/Pgi\text{-}2^b$; $Pgi\text{-}1^b/Pgi\text{-}1^b$, $Pgi\text{-}2^b/Pgi\text{-}2^b$; $Pgi\text{-}1^b/Pgi\text{-}1^c$, $Pgi\text{-}2^b/Pgi\text{-}2^b$; $Pgi\text{-}1^c/Pgi\text{-}1^c$, $Pgi\text{-}2^b/Pgi\text{-}2^b$; $Pgi\text{-}1^b/Pgi\text{-}1^b$, $Pgi\text{-}2^a/Pgi\text{-}2^b$; $Pgi\text{-}1^b/Pgi\text{-}1^c$, $Pgi\text{-}2^a/Pgi\text{-}2^b$; $Pgi\text{-}1^b/Pgi\text{-}1^b$, $Pgi\text{-}2^b/Pgi\text{-}2^b$; $Pgi\text{-}1^b/Pgi\text{-}1^b$, $Pgi\text{-}2^b/Pgi\text{-}2^c$; and $Pgi\text{-}1^b/Pgi\text{-}1^b$, $Pgi\text{-}2^b/Pgi\text{-}2^b$.

of subunits in the whole organism is probably not random. Alternatively, the heterodimer may be less stable *in vivo*.

Individuals heterozygous at the PGI loci exhibit phenotypic patterns fully consistent with the above model. Fish heterozygous at a single locus show six bands, representing all possible associations of three types of subunits as follows: AA, A'A, A'A', AB, A'B, and BB (see Figs. 1 and 2). Individuals heterozygous at both loci often show only nine bands, rather than the ten predicted by random association of four types of subunits, but most likely this is due to overlap in mobility of some of the isozymes.

Tissue Distribution

Patterns of tissue and organ distribution of the PGI enzymes were examined in 16 species belonging to 15 families of Osteichthyes (Table I). Blood, eye, eggs, brain, liver, kidney, heart, gonad, spleen, skeletal muscle, gills, intestine, gallbladder, and pyloric cecae were all examined in a few of the species, although muscle, heart, and liver proved to be most informative. The organs and tissues were homogenized separately and electrophoresed as described above. A typical pattern of tissue distribution of PGI is shown in Fig. 3.

Table I. Species of Fish Examined for Phosphoglucose Isomerase

Superorder Order	Family	Scientific name	Common name	Sample size	Probable No. loci	Individual tissues examined	Heterozygotes observed
Semionotiformes	Lepisosteidae	*Lepisosteus oculatus*	Spotted gar	1	1	Yes	No
		Lepisosteus platostomus	Shortnose gar	2	1	Yes	Yes
Amiiformes	Amiidae	*Amia calva*	Bowfin	1	1	Yes	No
Elopomorpha Anguilliformes	Congridae	*Congrina flava*	Yellow conger	1	2	Yes	No
Clupeomorpha Clupeiformes	Clupeidae	*Brevoortia tyrannus*	Atlantic menhaden	2	≥ 1	No	Yes
		Clupea harengus	Atlantic herring	2	≥ 1	No	Yes
		Dorosoma cepedianum	Gizzard shad	6	≥ 1	Yes	No
Protacanthopterygii Salmoniformes	Salmonidae	*Salmo gairdneri*	Rainbow trout	30	3	Yes	No
		Salmo trutta	Brown trout	5	3	Yes	No
		Salvelinus fontinalis	Brook trout	5	3	No	No
	Esocidae	*Esox americanus*	Grass pickerel	2	2	No	No
Myctophiformes	Synodontidae	*Synodus foetens*	Inshore lizardfish	1	≥ 1	No	No
Ostariophysi Cypriniformes	Characidae	*Astyanax mexicanus*	Mexican tetra	400	2	Yes	Yes
	Cyprinidae	*Notropis venustus*	Blacktail shiner	2	2	No	Yes
		Carassius auratus	Goldfish	6	3	Yes	Yes
Siluriformes	Ictaluridae	*Ictalurus melas*	Black bullhead	1	2	No	No
	Ariidae	*Arius felis*	Sea catfish	2	2	Yes	Yes
Paracanthopterygii Batrachoidiformes	Batrachoididae	*Porichthys porosissimus*	Atlantic misdhipman	1	≥ 1	No	No
Gadiformes	Gadidae	*Microgadus tomcod*	Atlantic tomcod	1	2	Yes	Yes
	Ophidiidae	*Rissola marginata*	Striped cusk-eel	1	2	No	No
	Zoarcidae		Eelpout	1	2	Yes	No

Order	Family	Species	Common name				
Atherinomorpha	Cyprinodontidae	*Cyprinodon variegatus*	Sheepshead minnow	2	2	No	No
		Fundulus heteroclitus	Mummichog	1	2	No	No
	Poeciliidae	*Gambusia geiseri*	Largespring gambusia	15	2	No	No
		Gambusia affinis	Mosquitofish	15	2	No	Yes
		Gambusia heterochir	Clear Creek gambusia	1	2	No	Yes
		Poecilia formosa	Amazon molly	3	2	No	No
	Atherinidae	*Menidia menidia*	Atlantic silverside	2	1	No	Yes
Acanthopterygii							
Gasterosteiformes	Syngnathidae	*Syngnathus fuscus*	Northern pipefish	1	2	No	No
Perciformes	Serranidae	*Centropristis striata*	Black sea bass	1	2	Yes	No
	Centrarchidae	*Lepomis cyanellus*	Green sunfish	30	2	No	Yes
		Lepomis macrochirus	Bluegill	2000	2	No	No
		Lepomis humilis	Orangespotted sunfish	40	2	No	Yes
	Percidae	*Micropterus salmoides*	Largemouth bass	5	2	No	No
		Percina caprodes	Logperch	1	2	No	No
	Pomatomidae	*Pomatomus saltatrix*	Bluefish	1	≥ 1	Yes	No
	Pomadasyidae	*Orthopristis chrysoptera*	Pigfish	1	2	Yes	No
	Sparidae	*Stenotomus chrysops*	Scup	1	2	Yes	No
	Sciaenidae	*Cyoscion nothus*	Silver seatrout	1	2	No	No
		Micropogon undulatus	Atlantic croaker	1	2	No	No
	Cichlidae	*Cichlasoma cyanoguttatum*	Rio Grande perch	1	2	No	No
	Labridae	*Tautoga onitis*	Tautog	1	2	Yes	No
		Tautogolabrus adspersus	Cunner	1	2	No	No
	Mugilidae	*Mugil cephalus*	Striped mullet	2	2	Yes	No
	Trichiuridae	*Trichiurus lepturus*	Atlantic cutlassfish	4	2	No	No
	Scombridae	*Scomber scombrus*	Atlantic mackerel	1	2	No	No
	Stromateidae	*Peprilus triacanthus*	Butterfish	2	2	No	No
	Triglidae	*Prionotus* sp.	Sea Robin	1	2	No	No
	Cottidae	*Hemitripterus americanus*	Sea raven	2	2	Yes	No
Pleuronectiformes	Bothidae	*Ancylopsetta quadrocellata*	Ocellated flounder	2	2	No	No
	Soleidae	*Achirus lineatus*	Lined sole	1	2	No	No
	Cynoglossidae	*Symphurus plagiusa*	Blackcheek tonguefish	1	≥ 1	No	Yes
Tetraodontiformes	Tetraodontidae	*Sphoeroides parvus*	Least puffer	1	2	No	Yes

135

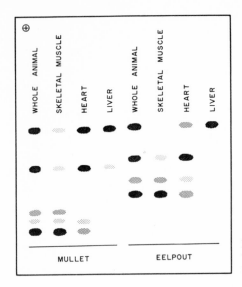

Fig. 3. Diagrammatic representation of tissue distribution of phosphoglucose isomerases in typical teleosts.

In all species possessing two PGI loci (with the exception of the yellow conger, *Congrina flava*), the slower-migrating PGI-2 homodimer is strongly expressed in skeletal muscle and is weak or absent in other tissues. PGI-1 is very weak in muscle. The heterodimer is strongly expressed in the heart, often along with moderate PGI-1 and PGI-2 activity. All other tissues and organs predominantly exhibit PGI-1. In the yellow conger, *C. flava*, white muscle and eggs express both homodimer bands, while heart and liver show mainly PGI-1.

Characterization of PGI in *A. mexicanus*

In order to substantiate more fully kinetic and molecular differences of the PGI enzymes and hence their genetic control by independent gene loci, we chose to examine in more detail the properties of the PGI homodimers in the characid teleost *A. mexicanus*. Both PGI loci are polymorphic in *A. mexicanus*. Populations along a 50 mile transect in northeastern Mexico exhibit eight alleles at the *Pgi-1* locus and six alleles at the *Pgi-2* locus. The high frequency of heterozygotes, the close correspondence of genotypes at both loci with Hardy–Weinberg expectations, and the occurrence of almost all possible combinations of alleles at the two loci in the populations sampled provide additional evidence for the presence and independence of two genetic loci.

The homodimeric forms of PGI were isolated from specimens of *A. mexicanus* as described above. Heat inactivation experiments demonstrate

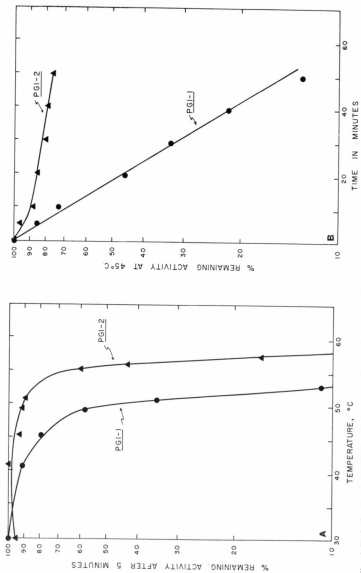

Fig. 4. Relative heat stabilities of phosphoglucose isomerase homodimers from *A. mexicanus*. A: Heating at different temperatures for 5 min. B: Heating at 45 C for different periods.

137

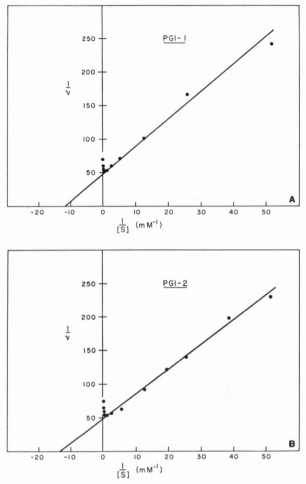

Fig. 5. Double reciprocal plots of *A. mexicanus* phosphoglucose isomerase homodimers with fructose 6-phosphate as varied substrate. A: PGI-1. B: PGI-2.

striking differences in the thermal stabilities of the two forms (Fig. 4). PGI-1 and PGI-2 lose 90% and 23% of their activity, respectively, after being heated at 45 C for 50 min. Furthermore, all PGI-1 activity is lost after 5 min of heating at 53 C, whereas 85% of PGI-2 activity remains.

Michaelis constants for the PGI homodimers were determined from Lineweaver–Burk reciprocal plots, using noninhibitory ranges of F6P concentration (Fig. 5). The K_m values for F6P are almost identical in PGI-1 and PGI-2: 7.0×10^{-5} and 7.6×10^{-5}, respectively. These values are com-

138

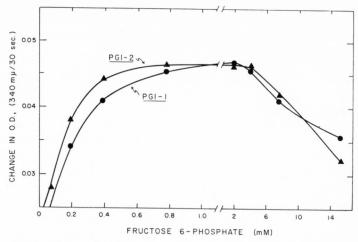

Fig. 6. Substrate inhibition of *A. mexicanus* phosphoglucose isomerase homodimers by fructose 6-phosphate.

parable to those found for the enzymes from bovine mammary gland (Hines and Wolfe, 1963), rabbit brain and skeletal muscle, and human erythrocytes (Kahana *et al.*, 1960). Both PGI-1 and PGI-2 are inhibited by high F6P substrate concentrations (Fig. 6).

The action of isomerase inhibitors on the activity of PGI homodimers from *Astyanax* extracts is shown in Table II. Phosphoenol pyruvate (PEP) and adenosine 5′-triphosphate (ATP) inhibit both PGI forms, but 6-phosphogluconate apparently has little effect at concentration 1×10^{-4} M. The inhibitory effect of higher concentrations of ATP appears to be greater on PGI-1 than on PGI-2.

Table II. Effect of Inhibitors on Phosphoglucose Isomerase Homodimers Isolated from *A. mexicanus*

Inhibitor	Concentration (*M*)	Percent Inhibition of PGI-1	PGI-2
Phosphoenol pyruvate	1×10^{-3}	8	21
	2.5×10^{-3}	46	46
	5×10^{-3}	90	98
Adenosine 5′-triphosphate	1×10^{-3}	3	10
	2.5×10^{-3}	13	22
	4×10^{-3}	32	30
	4.5×10^{-3}	38	33
	5×10^{-3}	64	33
6-Phosphogluconate	1×10^{-4}	4	10

DISCUSSION

Homology and Divergent Specialization of PGI Loci

The formation of a heterodimeric enzyme by the association of subunits encoded by the two PGI loci provides strong evidence for their homology and hence for the origin of one of the loci through duplication. The well-known gene duplications for malate dehydrogenase, lactate dehydrogenase, and hemoglobins (Kitto and Kaplan, 1966; Fondy and Kaplan, 1965; Kaplan, 1965; Ingram, 1961) provide analogous examples. These genes also encode subunits which can associate to form functional proteins. Furthermore, the PGI homodimeric enzymes in *A. mexicanus* have certain similar kinetic properties such as Michaelis constants and reaction to inhibitors (phosphoenol pyruvate and 6-phosphogluconate). It is unlikely that the similarities of these molecules and their ability to form heterodimers (particularly in the heart) are the result of convergence in coding properties of nonhomologous genetic loci.

The evolutionary potential of duplicate genes was appreciated before molecular and biochemical techniques had fully documented their existence. Huxley's (1942) concept of divergent specialization of duplicate genetic loci giving great delicacy of adjustment to the genome was soon extended by Lewis (1951) and Stephens (1951) to a realization of the dual dynamic and conservative aspects of gene duplications. According to their view, gene duplications may initially release one locus from selection while the other maintains its previous function. The new locus is then free to undergo modifications through chromosomal rearrangements and mutations to assume new functions, frequently closely similar to those of the original gene. Our present views of gene duplications, supported with numerous examples and with a better understanding of the cellular mechanisms involved, represent only refinements of these original ideas (Watts and Watts, 1968*a*, Ohno, 1970).

The phosphoglucose isomerase enzymes encoded by the two independent genetic loci in *A. mexicanus* exemplify a modification and divergence in structure and function following gene duplication. The two enzymes, catalysing the same reaction, show striking differences in electrophoretic mobility, in heat inactivation properties, and probably also in reaction to the inhibitor adenosine 5′-triphosphate.

The specific pattern of tissue distribution is indirect evidence of biologically meaningful divergence in the PGI enzymes. PGI-2 is expressed very strongly in skeletal muscle but not elsewhere; the hybrid molecule is strongest in the heart; and PGI-1 predominates in most other tissues and organs. We are not yet in a position to discuss the physiological significance of the differential expression of subunits in the various tissues.

140

Evolutionary History of the PGI Gene Duplication in Osteichthyes

In an attempt to assess the extent of occurrence of two PGI loci in fish, with the ultimate aim of understanding the evolutionary history of the gene duplication and its time of origin, we have surveyed 53 species of bony fishes belonging to 38 families and 15 orders for the number of genetic loci encoding PGI. These species are listed in Table I, in natural or phyletic series as proposed by Bailey *et al.* (1970). In larger individuals of various species, the tissue distribution of the PGI enzymes was examined (Table I). In smaller specimens, the whole animal was ground for electrophoresis. Table I also indicates, for each species, whether any of the individuals examined were heterozygous at a PGI locus. The following discussion is based on Romer's (1966) phylogeny of bony fishes.

The duplicated gene for PGI is of widespread occurrence in fishes. For example, in the order Perciformes, which includes the great majority of spiny-rayed fishes (Lagler *et al.*, 1962), 19 of the 20 species examined belonging to 15 families unquestionably possess two PGI loci. Furthermore, in at least 38 of 46 species representing 34 families in the main line of teleost evolution (superorder Protacanthopterygii and derived forms), two loci are again present. Figure 7, modified from Romer (1966), summarizes the presumed family tree of higher bony fishes (Actinopterygii). A comparison of Fig. 7 and Table I confirms that representative species of all five superorders

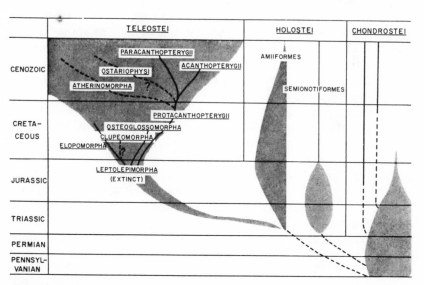

Fig. 7. "Family tree" of higher bony fishes (Actinopterygii). (Modified from Romer, 1966.)

141

of teleosts including and derived from the basal primitive stock Salmoniformes (Protacanthopterygii, Ostariophysi, Atherinomorpha, Paracanthopterygii, and Acanthopterygii) possess duplicated PGI loci. These five superorders include the great majority of all species of bony fishes and, for convenience, are specifically referred to as "higher teleosts" in the following discussion. [Recently, Nelson (1969) has reclassified the Protacanthopterygii, Paracanthopterygii, and Acanthopterygii into the cohort Euteleostei. A new superorder, Neoteleostei, now includes Myctophiformes, Paracanthopterygii, and Acanthopterygii. Or, according to Rosen and Patterson (1969), the Myctophiformes are elevated to superordinal status along with the Paracenthopterygii and Acanthopterygii. Since none of our conclusions are affected by these interpretations, we have chosen to retain Romer's classification until the matter is settled].

The patterns of tissue distribution of the PGI subunits of the higher teleost species and even their relative position on the gels show striking regularity. In all higher teleosts, PGI-2, always the slower-migrating, less anodal band, is invariably predominant in skeletal muscle, while the faster-migrating PGI-1 prevails in liver and other internal organs. The heart shows various concentrations of enzymes, but usually with the heterodimer or PGI-1 homodimer most strongly expressed. One or two satellite bands of very characteristic appearance consistently migrate slightly anodal to PGI-2.

In five of the 46 species of higher teleosts examined, evidence for genetic control of PGI by multiple loci is poor or lacking. In the lizard-fish and Atlantic midshipman, two distinct bands of similar mobility appear on gels, while in the bluefish and tonguefish only a single distinct band is present. The northern pipefish exhibits three bands very close in mobility, but since only a single specimen was available, the possibility that this phenotypic pattern results from heterozygosity at a single locus cannot be eliminated. The phenotypic patterns on the gels of these and other species clearly illustrate occasional difficulties of the electrophoretic technique. Two independent genetic loci encoding subunits of identical electrophoretic mobility would normally be scored as a single locus. The genetic basis of two or even three bands of similar mobility is likewise unclear, unless individual tissues are examined or occasional heterozygotes are seen. However, the following lines of evidence are advanced in favor of the proposal that these five species may also possess duplicate PGI loci along with all other higher teleosts: (1) the range of relative electrophoretic mobilities of PGI-1 and PGI-2 in other teleosts is considerable (great separation in some species, little in others), and, by chance, in certain species, we would expect PGI-1 and PGI-2 to exhibit identical or nearly identical mobilities; (2) the great majority of higher teleosts clearly exhibit two PGI loci, indicating early phylogenetic origin of the gene duplication; and (3) in the case of the bluefish and tonguefish, numerous other closely related species (belonging to the same orders) have two PGI loci.

Therefore, while the possibility of secondary loss of a duplicate PGI locus in a few higher teleosts cannot be excluded, we conclude that the original gene duplication occurred sometime before the branching of the Protacanthopterygii and is now manifested in most if not all higher teleosts. Moreover, the striking similarity of tissue distribution and relative electrophoretic mobilities of the homodimers in all higher teleosts examined strongly suggests that the two loci had already diverged in structure and function in an ancestor of the Protacanthopterygii. In other words, in this early ancestor, PGI-1 may have already assumed its primary function in skeletal muscle and PGI-2 in the other internal organs.

The Protacanthopterygii (from which probably arose all higher teleost orders), appear to have their origin in primitive salmon-like fish of the Upper Cretaceous, and in the pike (*Esox*) of the early Tertiary. However, by the Lower Cretaceous, at least three other primitive types of teleosts, giving rise to independent lines represented by several living forms, had also made their appearance. These were the superorders Elopomorpha, Osteoglossomorpha, and Clupeomorpha, comprising the remainder of the living teleosts.

The superorder Elopomorpha is today represented by the tarpon group of fishes and by eels (Anguilliformes). The only member of Elopomorpha examined in this study was the yellow conger, *C. flava*. Again, two PGI loci are demonstrable, but the pattern of tissue distribution appears to differ from that of higher teleosts. Skeletal muscle shows approximately equal concentrations of subunits encoded by each locus, as do the eggs, while heart shows only the PGI-1 homodimer. The heterodimer band is very weak in these tissues. Living representatives of the tropical superorder Osteoglossomorpha were not obtained for our studies.

Far more successful in modern times are the Clupeiformes, living representatives of a group, Clupeomorpha, whose origin is still uncertain. Early herring-like ancestors were well established in the Lower Cretaceous, but whether they arose independently from an ancestral teleost stock or are an offshoot of the elopomorphs is unknown. Three species of Clupeidae (menhaden, herring, and gizzard shad) show similar patterns on the gels. In each case, a single band stains intensely, with much lighter, slightly anodal satellite bands. This pattern appears similar to the typical banding pattern of the PGI-2 enzymes in all other teleosts. In addition, thin light bands, unlike the typical PGI-1 band of other teleosts, appear far anodal and cathodal to the intense band. Phenotypes of various tissues of the gizzard shad are closely similar. Whether these results are evidence for a single PGI locus or whether multiple loci are present which encode non-tissue-specific enzymes is uncertain. In either case, however, it appears that, relative to other teleosts, the PGI enzymes in the Clupeidae are under a different type of genetic control.

All species discussed to this point belong to the Teleostei, an assemblage that began flourishing in the Cretaceous and includes the great majority of living fishes. However, in the Jurassic and Lower Cretaceous, another group of more primitive bony fishes, the holosteans, were dominant. Although holosteans were almost completely replaced by the teleosts in the Upper Cretaceous, two genera survived and have representatives living today—the gars and bowfins. The bowfin (*Amia calva*) traces its ancestry back to the Triassic, where the Amiiformes diverged from the line leading to the teleosts (Fig. 7). The Semionotiformes, represented only by the living gars, probably diverged still earlier from a primitive stock, perhaps in the late Paleozoic.

The bowfin and two species of gar were examined for PGI activity. In each case, a single intense band plus the characteristic anodal secondary bands appear on the gel, in extracts both from whole animals and from various tissues. Thus there is no good evidence for more than a single genetic locus encoding PGI.

In summarizing this section, we may note that there are general trends observable from Table I which warrant an estimate of the time of origin of the gene duplication and a consideration of the general history of the divergent specialization of the PGI homodimers into various tissues. Assuming that holosteans have only a single PGI locus and have not secondarily lost a duplicate gene, the gene duplication must have arisen sometime after the phylogenetic branching of the holosteans and before the branching of the higher teleosts from the Elopomorpha. A likely candidate for this origin is Leptolepiformes, a generalized group prominent in the Jurassic which gave rise to all modern teleosts. Furthermore, the enzymes probably diverged in function (as evidenced by tissue specificities) very early in the history of teleosts, although probably after the Clupeiformes had branched from ances ral stock. It should be most interesting to study representatives of the Osteoglossomorpha to test these hypotheses. The chondrosteans, which were the precursors of the holosteans and are represented by two living American families, the sturgeons (Acipenseridae) and paddlefishes (Polyodontidae), also remain to be examined.

Gene Duplication Through Polyploidization

Gene duplications may arise through various mechanisms of tandem duplication involving only small portions of the genome or through polyploidization. Ohno (1970) and Ohno *et al.* (1968) view the two mechanisms as being complementary and further feel that, since polyploid evolution is denied to mammals, birds, and reptiles, due to their well-established chromosomal sex determining mechanisms, most successful achievements of gene duplications in vertebrates must have taken place at the stage of fishes and amphibians.

Multiple loci, presumably resulting from duplications of genes encoding a wide variety of proteins, have previously been reported in fishes: 6-phosphogluconate dehydrogenase in certain cyprinids (Klose *et al.*, 1969; Bender and Ohno, 1968), hemoglobins in hagfish and salmonids (Ohno and Morrison, 1966; Tsuyuki and Gadd, 1963), malate dehydrogenase in salmonids (Bailey *et al.*, 1969, 1970), lactate dehydrogenase in cyprinids and salmonids (Klose

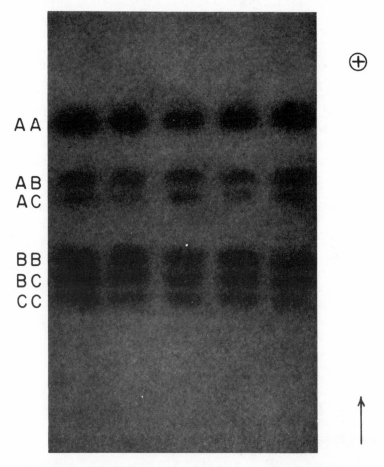

Fig. 8. Phenotypes of phosphoglucose isomerases from whole-animal extracts of brook trout. All individuals are presumably homozygous at three PGI loci. A, B, and C: Subunits encoded by *Pgi-1*, *Pgi-2*, and *Pgi-3*, respectively.

et al., 1968, 1969; Massaro and Markert, 1968; Morrison and Wright, 1966), and isocitrate dehydrogenase in certain cyprinids (Quiroz-Gutierrez and Ohno, 1970). Several of these gene duplications are apparently restricted to a few salmonids and cyprinids which, from cytological and DNA evidence, are probably tetraploid (Ohno *et al.*, 1967; Massaro and Markert, 1968; Ohno, 1970).

Three presumably tetraploid salmonid species (brook trout, brown trout, and rainbow trout) were examined for PGI activity. All individuals exhibit six bands, most likely representing random association of subunits encoded by three genetic loci (Fig. 8). Apparently, at least one of the original pair of PGI loci has reduplicated to form a third locus as a result of polyploidization. Furthermore, PGI-3 (the homodimer produced by the third locus) resembles PGI-2 in tissue distribution (predominant in muscle, weak elsewhere) and in electrophoretic mobility. Hence, as a result of polyploidization, two former *Pgi-2* locus alleles have been incorporated into the genome. However, in salmonids, the *Pgi-1* locus shows no evidence of having reduplicated. Perhaps alleles of the *Pgi-1* locus incorporated into the genome are identical, and thus the organism simply has more loci making PGI-1, or perhaps the duplicated *Pgi-1* locus did not confer a selective advantage on the organism and thus was silenced.

In tetraploid goldfish (Cyprinidae), six major bands are present in all individuals, again giving evidence for three genetic loci. In this case, duplication of the *Pgi-1* locus through polyploidization has resulted in a new locus encoding a homodimer (PGI-4) which, along with PGI-1 and their heterodimer, predominates in liver and other organs and is very weak in skeletal muscle. The heart expresses PGI-1, PGI-2, PGI-4, and their heterodimers, although the PGI-1–PGI-4 complex is strongest. PGI-2 is, as usual, strongest in muscle.

Tetraploid trout and goldfish demonstrate that new genetic loci can be incorporated into the genome through polyploidization, although not all genetic loci give rise to new loci detectable by electrophoresis. Unfortunately, we do not know whether the original PGI gene duplication in fishes arose through tandem duplication or through polyploidization.

ACKNOWLEDGMENTS

Part of this work was carried out in the laboratories of Dr. Robert K. Selander and Dr. Michael H. Smith, to whom we are indebted for invaluable support and assistance. We also thank James Lemberg and Bob Gardner for technical assistance.

REFERENCES

Avise, J. C., and Selander, R. K. (1972). Evolutionary genetics of cave-dwelling fishes of the genus *Astyanax*. *Evolution* **26**:1.

Bailey, G. S., Cock, G. T., and Wilson, A. C. (1969). Gene duplication in fishes: Malate dehydrogenase of salmon and trout. *Biochem. Biophys. Res. Commun.* **34**:605.

Bailey, G. S., Wilson, A. C., Halver, J. E., and Johnson, C. L. (1970). Multiple forms of supernatant malate dehydrogenase in salmonid fishes. *J. Biol. Chem.* **245**:5927.

Bailey, R. M., Fitch, J. E., Herald, E. S., Lachner, E. A., Lindsey, C. C., Robins, C. R., and Scott, W. B. (1970). A list of common and scientific names of fishes from the United States and Canada. *Am. Fish. Soc. Serial. Publ.*, No. 6.

Bender, K., and Ohno, S. (1968). Duplication of the autosomally inherited 6-phosphogluconate dehydrogenase gene locus in tetraploid species of cyprinid fish. *Biochem. Genet.* **2**:101.

Carter, N. D., and Parr, C. W. (1967). Isoenzymes of phosphoglucose isomerase in mice. *Nature* **216**:511.

DeLorenzo, R. J., and Ruddle, F. H. (1969). Genetic control of two electrophoretic variants of glucose phosphate isomerase in the mouse (*Mus musculus*). *Biochem. Genet.* **3**:151.

Detter, J. C., Ways, P. O., Giblett, E. R., Baughman, M. A., Hopkinson, D. A., Povey, S., and Harris, H. (1968). Inherited variations in human phosphohexose isomerase. *Ann. Hum. Genet.* **31**:329.

Fondy, T. P., and Kaplan, N. O. (1965). Structural and functional properties of the H and M subunits of lactic dehydrogenases. *Ann. N.Y. Acad. Sci.* **119**:888.

Hines, M. C., and Wolfe, R. G. (1963). Phosphoglucose isomerase. II. Influence of pH on kinetic parameters. *Biochemistry* **2**:770.

Huxley, J. (1942). *Evolution: The Modern Synthesis*, Allen and Unwin, London.

Ingram, V. M. (1961). Gene evolution and the haemoglobins. *Nature* **189**:704.

Johnson, W. E., and Selander, R. K. (1971). Protein variation and systematics in kangaroo rats (genus *Dipodomys*). *Syst. Zool.* **20**: 377.

Johnson, W. E., Selander, R. K., Smith, M. H., and Kim, Y. J. (1972). Biochemical genetics of sibling species of the cotton rat (*Sigmodon*). *Stud. Genet.* (*Univ. Texas Publ.*) **7** (7213):297.

Kahana, S. E., Lowry, O. H., Schulz, D. W., Passonneau, J. V., and Crawford, E. J. (1960). The kinetics of phosphoglucoisomerase. *J. Biol. Chem.* **235**: 2178.

Kaplan, N. O. (1965). Evolution of dehydrogenases. In Bryson, V., and Vogel, H. J. (eds.), *Evolving Genes and Proteins*, Academic Press, New York, p.243.

Kitto, G. B., and Kaplan, N. O. (1966). Purification and properties of chicken heart mitochondrial and supernatant malic dehydrogenases. *Biochemistry* **5**:3966.

Klose, J., Wolf, U., Hitzeroth, H., Ritter, H., Atkin, N. B., and Ohno, S. (1968). Duplication of LDH gene loci by polyploidization in the fish order Clupeiformes. *Humangenetik* **5**:190.

Klose, J., Wolf, U., Hitzeroth, H., Ritter, H., and Ohno, S. (1969). Polyploidization in the fish family Cyprinidae, order Cypriniformes. II. Duplication of the gene loci coding for lactate dehydrogenase (E.C.:1.1.1.27) and 6-phosphogluconate dehydrogenase (E.C.:1.1.1.44) in various species of Cyprinidae. *Humangenetik* **7**:245.

Lagler, K. F., Bardach, J. E., and Miller, R. R. (1962). *Ichthyology*, Wiley, New York.

Lewis, E. B. (1951). Pseudoallelism and gene evolution. *Cold Spring Harbor Symp. Quant. Biol.* **16**:159.

Massaro, E. J., and Markert, C. L. (1968). Isozyme patterns of salmonoid fishes: Evidences for multiple cistrons for lactate dehydrogenase polypeptides. *J. Exptl. Zool.* **168**: 223.

Morrison, W. J., and Wright, J. E. (1966). Genetic analysis of three lactate dehydrogenase isozyme system in trout: Evidence for linkage for genes coding subunits A and B. *J. Exptl. Zool.* **163**:259.

Nakagawa, Y., and Noltmann, E. A. (1967). Multiple forms of yeast phosphoglucose isomerase. I. Resolution of the crystalline enzyme into three isoenzymes. *J. Biol. Chem.* **20**:4782.

Nelson, G. J. (1969). Gill arches and the phylogeny of fishes, with notes on the classification of vertebrates. *Bull. Am. Mus. Nat. Hist.* **141**:475.

Noltmann, E. A. (1964). Isolation of crystalline phosphoglucose isomerase from rabbit muscle. *J. Biol. Chem.* **239**:1545.

Nottebohm, F., and Selander, R. K. (1972). Vocal dialects and gene frequencies in the Chingolo sparrow (*Zonotrichia capensis*). *Condor* **74**:137.

Ohno, S. (1970). *Evolution by Gene Duplication*, Springer-Verlag, New York.

Ohno, S., and Morrison, M. (1966). Multiple gene loci for the monomeric hemoglobin of the hagfish (*Eptatretus stoutii*). *Science* **154**:1034.

Ohno, S., Muramoto, J., Christian, L., and Atkin, N. B. (1967). Diploid–tetraploid relationship among Old-World members of the fish family Cyprinidae. *Chromosoma* **23**:1.

Ohno, S., Wolf, U., and Atkin, N. B. (1968). Evolution from fish to mammals by gene duplication. *Hereditas* **59**:169.

Prasad, R., and Wright, D. A. (1971). Sponge method of preparative electrophoresis. *Isozyme Bull.* **4**:12.

Quiroz-Gutierrez, A., and Ohno, S. (1970). The evidence of gene duplication for S-form NADP-linked isocitrate dehydrogenase in carp and goldfish. *Biochem. Genet.* **4**:93.

Ralin, D. B., Selander, R. K., and Yang, S. Y. (1972). Protein variation and systematics in the genus *Hyla*. In preparation.

Ramsey, P. R., Avise, J. C., Smith, M. H., and Urbston, D. F. (1972). Protein variation and genic heterozygosity in white-tailed deer (*Odocoileus virginianus*). In preparation.

Romer, A. S. (1966). *Vertebrate Paleontology*, 3rd ed., University of Chicago Press, Chicago.

Rosen, D. E., and Patterson, C. (1969). The structure and relationships of the paracanthopterygian fishes. *Bull. Am. Mus. Nat. Hist.* **141**: 357.

Selander, R. K., Hunt, W. G., and Yang, S. Y. (1969). Protein polymorphism and genic heterozygosity in two European subspecies of the house mouse. *Evolution* **23**:379.

Selander, R. K., Yang, S. R., Lewontin, R. C., and Johnson, W. E. (1970). Genetic variation in the horseshoe crab (*Limulus polyphemus*), a phylogenetic "relic." *Evolution* **24**:402.

Selander, R. K., Smith, M. H., Yang, S. Y., Johnson, W. E., and Gentry, J. B. (1971). Biochemical polymorphism and systematics in the genus *Peromyscus*. I. Variation in the old-field mouse (*Peromyscus polionotus*). *Stud. Genet.* (*Univ. Texas Publ.*) **6 (7103)**:49.

Selander, R. K., Johnson, W. E., and Avise, J. C. (1972). Biochemical population genetics of fiddler crabs (*Uca*). In preparation.

Stephens, S. G. (1951). Possible significance of duplication in evolution. *Advan. Genet.* **4**:247.

Tsuyuki, H., and Gadd, R. E. A. (1963). The multiple hemoglobins of some members of the Salmonidae family. *Biochim. Biophys. Acta* **71**:219.

Watts, R. L., and Watts, D. C. (1968a). Gene duplication and the evolution of enzymes. *Nature* **217**:1125.

Watts, R. L., and Watts, D. C. (1968b). The implications for molecular evolution of possible mechanisms of primary gene duplication. *J. Theoret. Biol.* **20**:227.

Webster, T. P., Selander, R. K., and Yang, S. Y. (1972). Genetic variability and similarity in the *Anolis* lizards of Bimini. *Evolution* (submitted).

Whitt, G. S. (1970). Genetic variation of supernatant and mitochondrial malate dehydrogenase isozymes in the teleost *Fundulus heteroclitus*. *Experientia* **26**:734.

Yoshida, A., and Carter, N. D. (1969). Nature of rabbit phosphoglucose isomerase isozymes. *Biochim. Biophys. Acta* **194**:151.

Lens Protein Polymorphisms in Hatchery and Natural Populations of Brook Trout, *Salvelinus fontinalis* (Mitchill)[1]

LARRY R. ECKROAT

ABSTRACT

Lens protein phenotype and allele frequencies of three autosomal loci were analyzed by acrylamide gel electrophoresis for 1,164 specimens from four hatchery populations of brook trout, *Salvelinus fontinalis* (Mitchill), each representing combined progenies from random matings of hundreds of parents. A deficiency of heterozygous genotypes at one locus, according to Hardy-Weinberg analysis, may indicate a degree of inbreeding in the hatchery environment. Allele frequencies of the lens protein variations appeared to be relatively stable from year to year within a particular hatchery. Gene frequency analyses for the lens protein variations disclosed that brook trout populations from some of the different hatcheries were distinguishable from each other.

Analyses of the lens proteins for 547 specimens from nine natural populations of brook trout revealed no genetic divergence in these small isolated populations. A length-frequency analysis of one population indicated that allele frequencies were independent of length classes. Evidence was presented suggesting that a distance greater than 300 yards may act as an isolating mechanism leading to genetic divergence of a brook trout population in one small stream. Gene frequency analyses for the lens protein variations disclosed that some of the brook trout natural populations from different areas sampled were distinguishable from each other.

INTRODUCTION

The application of the principles of population genetics to the study of various polymorphic proteins has indicated that allele frequencies often are quite different among populations of the same species. In the past few years much attention has been directed toward such studies on the commercially important fishes. The objectives of these studies have been to find genetic polymorphisms and to identify separate breeding population units using the allele frequencies of these polymorphic systems. Usefulness of such genetic study was discussed by Marr (1957), Marr and Sprague (1963), and Fujino (1970).

The serum esterases of skipjack tuna, *Euthynnus pelamis* (Linnaeus), were reported to

[1] This investigation was supported in part by Grant GB4624 to J. E. Wright, Jr. from the National Science Foundation.

149

be applicable to the problem of identifying subpopulations or isolated breeding populations (Fujino and Kang, 1968b). The Y system of skipjack tuna blood groups also appeared useful in subpopulation identification; however, a statistical analysis of gene frequencies failed to demonstrate a significant heterogeneity between populations (Fujino and Kazama, 1968). In a study involving serum transferrins in three species of Atlantic and Pacific tunas, Fujino and Kang (1968a) found no significant heterogeneity of gene frequency distribution among different geographical areas or among different stocks in each area. The results of Barrett and Tsuyuki (1967) conflicted with the preceding report; they reported that differences in allele frequencies at the serum transferrin locus in tunas suggested a usefulness for population studies.

A clinal distribution of serum esterase allele frequencies in catostomid fish populations was suggested to be due to a selective maintenance of heterogeneity (Koehn and Rasmussen, 1967). Electrophoretic analyses of the serum proteins of white bass, *Morone chrysops* (Rafinesque), in several Wisconsin lake populations (Wright and Hasler, 1967), indicated that homing behavior and geographic distance are effective isolating mechanisms for this species.

Molecular differences in eye lens proteins of ocean whitefish, *Caulolatilus princeps* (Jenyns), identified by cellulose acetate electrophoresis by Smith and Goldstein (1967) apparently reflected intra-specific genetic variation. Their evidence suggested that the eye lens proteins could be used to distinguish separate breeding populations. The lens proteins of five scombroids were examined by micro-starch gel electrophoresis (Barrett and Williams, 1967). Intra-specific variation was found in Pacific bonito, *Sarda chiliensis* (Cuvier), but was interpreted to be mobility differences of a single fraction related to ontogenetic factors.

Calculations of interpopulation heterogeneities of allele frequency differences for both the LDH-B and the transferrin loci permitted

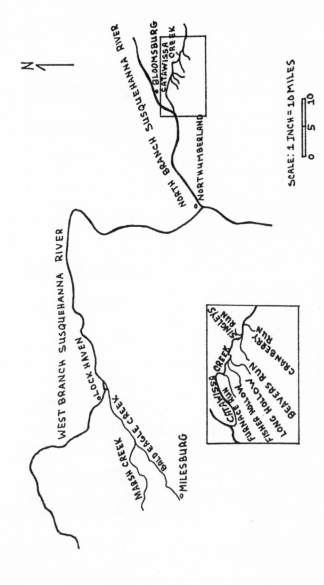

FIGURE 1.—Location of natural populations of brook trout sampled in Pennsylvania.

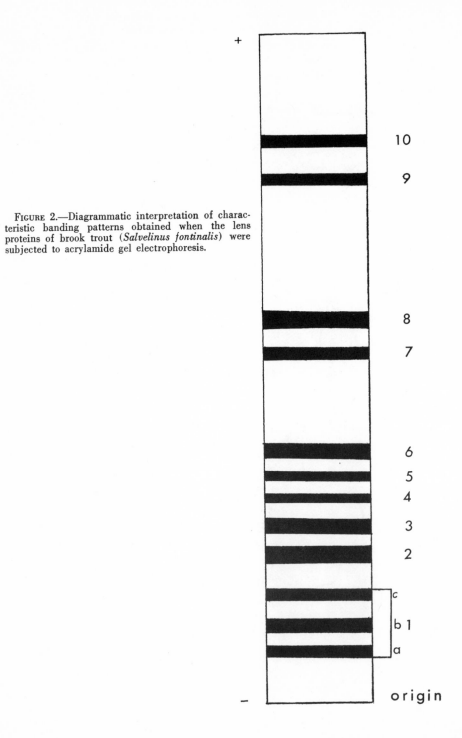

FIGURE 2.—Diagrammatic interpretation of characteristic banding patterns obtained when the lens proteins of brook trout (*Salvelinus fontinalis*) were subjected to acrylamide gel electrophoresis.

Wright and Atherton (1970) to distinguish the same hatchery populations and some of the same wild populations which were used in the present study.

Analyses of progenies of matings between individual brook trout of predetermined lens protein phenotypes revealed that two bands visualized on acrylamide gel electrophorograms are each controlled by two alleles at an autosomal locus and are inherited in a simple Mendelian manner (Eckroat and Wright, 1969). These two loci assort independently. Another band appears to be controlled by a dominant allele at an autosomal locus but this needs to be confirmed by analyses of F_2 progenies.

Since the genetic characteristics of some of the lens protein characters in brook trout were known, this study was undertaken with the following objectives: (1) to determine if the allele frequencies of the lens protein loci can be used to distinguish separate breeding population units; (2) to obtain the relative degree of lens protein polymorphisms in population units; and (3) to survey numerous individuals with the possibility of revealing new genetic markers.

MATERIALS AND METHODS

A total of 1,164 specimens of brook trout were obtained from live stocks reared at the Reynoldsdale Hatchery (Pennsylvania Fish Commission), the Edray and Ridge Hatcheries (West Virginia Game and Fish Division), and the Warren Hatchery (New Hampshire Department of Fish and Game). The samples were collected at random from among the hatchery production fish which represented progenies from random matings of hundreds of parents.

A total of 516 brook trout were sampled from natural populations in Centre and Columbia Counties, Pennsylvania, and 31 fish from Montana and Connecticut. The fish were taken by sportfishing and electrofishing. The approximate locations of the sources in Pennsylvania are indicated in Figure 1.

Lens samples were obtained by anesthetizing

153

the fish with 3-methyl-1 pentyn-3 ol (Leitritz, 1959). The lens was removed by making a crescent-shaped cut in the cornea with a dissecting knife and by applying a slight downward pressure with a pair of dissecting forceps. The lens was blotted quickly on absorbent paper toweling to remove foreign material, placed in a 6×50 mm tube under refrigeration, and used for analysis immediately or after storage at -20 C. Fingerling fish were killed and both lenses removed; however, with yearling fish only one lens was removed without killing the fish. All fish samples from natural populations were killed because other tissues (blood, liver, eye) were collected for other protein studies.

Lens samples were homogenized with a ground glass rod made to fit loosely into the storage tubes. The two lenses from fingerling trout were homogenized in 0.1 cc of glass distilled water, while the one lens from a yearling or adult fish was homogenized in 0.5 cc of glass distilled water. The homogenate was centrifuged at 4 C for 10 minutes at 1,000 g and the supernatant utilized for electrophoresis.

Electrophoresis in acrylamide gel was carried out with the Model 12 apparatus manufactured by Canalco Industrial Corporation, Rockville, Maryland. Procedures were similar to those of Ornstein (1964) and Davis (1964). A separating gel concentration of 7.5% acrylamide was used; the dimensions of the cylindrical columns were 65×5 mm. Electrophoresis was done at room temperature with a tris-glycine buffer (0.012 M tris and 0.095 M glycine), pH 8.3, and a regulated electric current of 5 ma per column. Gels were stained with 0.7% naphthol blue-black in 7.5% acetic acid for approximately 45 minutes and then destained electrophoretically in 7.5% acetic acid. A destaining apparatus, nine inches square, similar to the Canalco unit was constructed from 6 mm plexiglas. Holes were arranged in a circle so that 24 gels could be placed equi-distant from a platinum electrode.

154

TABLE 1.—*Eye lens phenotype and allele frequencies in hatchery brook trout populations*

		Phenotypes and dominant allele frequencies									
		Band 2				Band 4			Band 5		
Population		H	I	L	LP2	+	0	LP4	+	0	LP5
Pennsylvania:											
Reynoldsdale 2-yr-olds, 1968	Observed	177	25	9	.90	116	95	.33	3	208	.88
	Expected	171	19	21							
Reynoldsdale Fingerlings, 1968	Observed[1]	221	58	13	.86	137	155	.27	8	284	.83
	Expected	216	70	6							
New Hampshire:											
Warren (Gilbert) Yearlings, 1966	Observed[1]	21	5	4	.78	3	27	.05	3	27	.68
	Expected	18	10	2							
Warren (Gilbert) Fingerlings, 1968	Observed[1]	62	44	31	.61	5	132	.02	1	136	.97
	Expected	51	65	21							
West Virginia:											
Ridge Yearlings, 1966	Observed	9	10	10	.48	—	29	.00	—	29	1.00
	Expected	7	14	8							
Ridge Yearlings, 1967	Observed	5	12	5	.50	—	22	.00	—	22	1.00
	Expected	6	11	6							
Ridge Fingerlings, 1968	Observed[1]	64	77	85	.45	—	226	.00	—	226	1.00
	Expected	46	112	68							
Edray 2-yr-old, 1968	Observed[1]	119	68	30	.71	—	217	.00	—	217	1.00
	Expected	109	89	18							

[1] Populations not in equilibrium.

A total of ten distinct bands were visualized in hatchery brook trout lens proteins, as illustrated in Figure 2. They were numbered 1 through 10 in the ascending order of their anodal migration. Phenotypic variations were expressed as varying intensities of band 2: High (H), Intermediate (I) and Low (L) concentrations, and by the presence (+) or absence (–) of bands 4 and 5. The inheritance of these variations was discussed by Eckroat and Wright, 1969.

Allele Frequencies of Lens Proteins in Hatchery Populations

The frequencies of the phenotypes and the dominant alleles for three lens protein loci (bands 2, 4, and 5) in four hatchery populations of brook trout are shown in Table 1. Allele frequencies were calculated on the basis of a two-allele system. For the system expressed by varying intensities of band 2, Hardy-Weinberg expectations of the three phenotypes were compared with the observed values by the Chi-square test. The results indicated a marked deficiency of the heterozygotes in many populations, resulting in deviations from the condition of genetic equilibrium. Comparison of the year classes from the same hatcheries indicates that the gene frequencies are relatively stable from year to year.

Intra- and inter-hatchery heterogeneities were tested by Chi-square of contingency; the number of dominant alleles at each of the three lens protein loci were the variables compared in the various populations. The results (Table 2) indicated that intra-hatchery heterogeneity was nonsignificant for all except the Warren, New Hampshire samples. When Chi-square indicated that the year class samples were

TABLE 2.—*Heterogeneity tests summarized for allele frequencies of three lens protein loci in hatchery brook trout populations*

Hatchery population	χ^2	df	Probability
Intra-population			
Ridge (West Virginia)	0.21	4	>.90
Warren (New Hampshire)	6.62	2	<.05
Reynoldsdale (Pennsylvania)	1.38	2	>.50
Inter-population			
Ridge versus Warren	20.56	2	<.001
Ridge versus Reynoldsdale	216.30	2	<.001
Warren versus Reynoldsdale	72.64	2	<.001
Edray versus Ridge	17.28	2	<.001
Edray versus Warren	10.82	2	<.005
Edray versus Reynoldsdale	140.68	2	<.001

homogeneous they were pooled for the inter-hatchery comparisons. For inter-hatchery comparisons involving the Warren population, the 1968 sample was used since it was the largest in size. The inter-hatchery heterogeneity tests indicated that all populations were significantly different from each other. These results can be compared with results obtained from comparisons of the dominant allele frequencies in the various hatchery populations by means of 95 percent confidence intervals as shown in Figure 3. (These were computed by an approximate method of interpolation, Pearson and Hartley, 1954.) This method of comparison indicates that the Reynoldsdale population is significantly different from all other populations but it does not indicate significant differences among other hatchery populations. As with the Chi-square test, a significant intra-hatchery heterogeneity is indicated for the Warren, New Hampshire population.

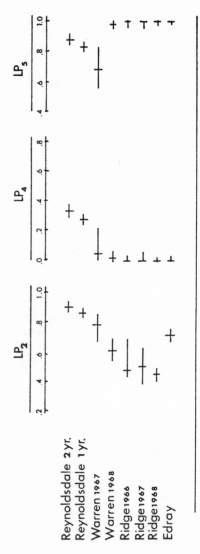

FIGURE 3.—Ninety-five percent confidence intervals for the allele frequencies of three lens protein loci in hatchery brook trout populations. Confidence intervals computed by an approximate method of interpolation (Pearson and Hartley, 1954). Actual allele frequencies are indicated by a vertical bar.

Frequencies of Lens Proteins Alleles in Natural Populations

Lens proteins of nine natural populations of brook trout were sampled in this study. Samples were collected from six streams which drain into Catawissa Creek, Columbia County, Pennsylvania. This creek has been polluted with acid mine drainage since the mid-1800's; as a result, no fish have been able to survive in this creek since that time. Thus, what prior to this pollution may have been an interbreeding population is now isolated into subpopulations. Frequencies of phenotypes and dominant alleles of the three lens protein loci for all natural population samples are shown in Table 3. It can be noted that all populations which showed segregation for band two are in Hardy-Weinberg equilibrium.

The allele frequencies were subjected to an inter-populational heterogeneity test and intra-populational tests were performed on the two Marsh Creek populations (Chi-square of contingency). These tests are summarized in Table 4. These can be compared with the results of inter-populational heterogeneity tests performed with 95% confidence intervals which indicate that the allele frequencies of the eye lens proteins of the brook trout can be used to distinguish some, but not all, populations. The 95% confidence intervals for the dominant allele frequencies of the various natural populations are shown in Figure 4. These intervals indicate that the Long Hollow population is significantly different from all others and that the Maiden Canyon, Montana, population is significantly different from the two Marsh Creek populations. All other inter-populational differences are not significant.

Intra-populational heterogeneity tests of the differences between two samples of the Marsh Creek population collected in different years and different areas of the stream showed the differences to be significant. This result indicates that geographic distance may cause the formation of subpopulations within a stream. To determine effects of stream distance as an isolating mechanism the 1968 Marsh Creek sample was collected in sequential areas of the

TABLE 3.—*Eye lens phenotypes and allele frequencies in natural brook trout populations*

Populations		Band 2				Band 4			Band 5		
		H	I	L	LP_2	+	O	LP_4	+	O	LP_5
Pennsylvania:											
Beaver Run	Observed	28	0	0	1.0	2	26	.04	0	28	1.0
	Expected	28									
Cranberry Run	Observed	55	0	0	1.0	8	47	.11	0	55	1.0
	Expected	55									
Fisher Hollow	Observed	20	2	0	.96	10	12	.26	0	22	1.0
	Expected	20	2								
Furnace Run	Observed	11	0	0	1.0	2	9	.10	0	11	1.0
	Expected	11									
Long Hollow	Observed	59	0	0	1.0	24	35	.23	7	52	.66
	Expected	59									
Singley Run	Observed	29	0	0	1.0	2	27	.04	0	29	1.0
	Expected	29									
Marsh Creek 1967	Observed	136	5	3	.96	64	80	.25	0	144	1.0
	Expected	133	11								
Marsh Creek 1968	Observed	165	3	0	.99	49	119	.16	0	168	1.0
	Expected	165	3								
Montana:											
Maiden Canyon	Observed	18	4	0	.91	0	22	.0	0	22	1.0
	Expected	18	4								
Connecticut:											
Squabble Brook	Observed	9	0	0	1.0	3	6	.18	0	9	1.0
	Expected	9	0								

160

TABLE 4.—*Heterogeneity test summarized for allele frequencies of three lens protein loci (bands 2, 4 and 5) in natural brook trout populations*

Population	Beaver Run	Cranberry Run	Fisher Hollow	Furnace Run	Long Hollow	Singley Run	1967 Creek Marsh	1968 Marsh Creek	Maiden Canyon	Squabble Brook
Beaver Run	–									
Cranberry Run	N.S.[1]	–								
Fisher Hollow	A	N.S.	–							
Furnace Run	B	B	N.S.	–						
Long Hollow	N.S.	N.S.	B	N.S.	–					
Singley Run	B	A	N.S.	N.S.	A	–				
1967 Marsh Creek	N.S.	N.S.	N.S.	N.S.	N.S.	A	–			
1968 Marsh Creek	N.S.	N.S.	B	N.S.	C	B	D	–		
Maiden Canyon	N.S.	N.S.	N.S.	N.S.	N.S.	N.S.	B	A	–	
Squabble Brook	N.S.	N.S.	N.S.	N.S.	N.S.	N.S.	N.S.	N.S.	A	–

[1] N.S. = No significant inter-populational heterogeneity.
A = Significant inter-populational heterogeneity at $P < .05$.
B = Significant inter-populational heterogeneity at $P < .01$.
C = Significant inter-populational heterogeneity at $P < .001$.
D = Significant inter-populational heterogeneity at $P < .05$.

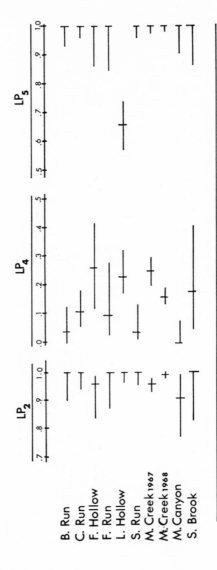

FIGURE 4.—Ninety-five percent confidence intervals for the allele frequencies of three lens protein loci in natural brook trout populations. Confidence intervals computed by an approximate method of interpolation (Pearson and Hartley, 1954). Actual allele frequencies are indicated by a vertical bar.

stream. The lengths of these areas were selected so as to coincide with natural boundaries such as logs, rock ledges, pools, and so forth. Analysis of these subsamples based on area suggested that within a distance of approximately 300 yards the brook trout population was of a homogeneous nature. The two main samples from Marsh Creek were collected in areas approximately 800 yards apart. Thus, based on lens protein allele differences, if geographic distance is an effective isolating barrier for brook trout in small streams it is at a distance of greater than 300 yards. Geographic distance was shown to be an effective isolating barrier in white bass populations in Wisconsin lakes (Wright and Hasler, 1967).

When the lengths of individual fish in the 1968 Marsh Creek sample were plotted against frequency, the population was divided into four distinct groups. This length-frequency method is often used as an approximate means of age determination. The frequencies of the dominant alleles for the three lens protein loci were subjected to an interaction Chi-square analysis which indicated that allele frequencies were independent of length classes.

During the survey of natural populations of brook trout additional bands were resolved. These bands were never observed in hatchery fish and did not occur in all fish from natural populations. The additional bands were designated as 6a, 8a, and 8b and are shown in Figure 5. Crosses were made between a native brook trout with band 6a present and a hatchery brook trout with band 6a absent. Analysis of the F_1 progenies of these crosses indicated that the presence of band 6a is controlled by a dominant allele at an autosomal locus. All of the F_1 progenies had band 6a present which would indicate that the native brook trout parent was homozygous (LP_{6a} LP_{6a}) for the dominant allele. It is quite surprising that both native fish would be homozygous dominant since the frequency of this band is relatively low in the natural populations. The appearance of this band, on the electrophorogram, is very dependent on the concentration of protein in the sample used. When a homogenate of one lens from a year-

163

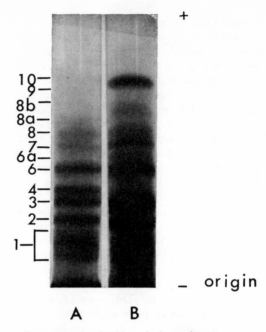

FIGURE 5.—Acrylamide gel electrophoretic separations in lens homogenates from two brook trout collected from a natural population (Marsh Creek) showing bands 6a, 8a and 8b. These bands were never found in hatchery brook trout. Trout A shows a phenotypic presence of bands 8a and 8b. Trout B shows a phenotypic presence of bands 6a and 8b.

ling fish or both lenses from a fingerling fish was used, band 6a was not always visible, but when a homogenate of both lenses from a yearling fish was used band 6a was clearly visible in all samples. In some of the F_1 progenies there appeared to be two very close bands instead of the usual one. Definite genetic evidence will require the making of an F_2 cross and analyzing the F_2 progenies. No matings were made involving bands 8a and 8b.

Lack of significant intra-hatchery heterogeneity at the Ridge and Reynoldsdale hatcheries indicates that samples from the year classes are of a homogeneous nature. This would seem to indicate that a genetic equilibrium ($\Delta q = 0$) has been reached in these hatchery populations; it would, however, require samples covering a longer period of time to make a definite statement. The fact that the Warren hatchery samples are heterogeneous may indicate that $\Delta q \neq 0$ or it could be caused by chance due to the relatively small sample size of the 1966 sample. Comparisons of the same two Warren samples indicated that alleles for LDH and serum transferrin loci are homogeneous (Wright and Atherton, 1970).

Significant inter-hatchery heterogeneity (Chi-square) indicates that every hatchery population is distinguishable by differences in the allele frequencies of the three lens protein loci. This result agrees with suggestions of Smith and Goldstein (1967) that lens protein polymorphisms of ocean whitefish can reflect differences between population units.

Comparisons of the dominant allele frequencies in the various hatchery and natural populations by means of 95% confidence intervals revealed that some populations could be distinguished by lens protein allele frequencies but that not as many were distinguishable as with heterogeneity tests (interaction Chi-square). This is because interaction Chi-square computed by contingency tables can be used to analyze enumeration data classified according to one or more variables while confidence intervals can be used to analyze enumeration data classified according to only one variable. Interaction Chi-square can be used to answer two questions: (1) is there dependence or independence of variables? and, (2) is there constancy of the set of ratios associated with one variable of classification irrespective of the categories of the other variables? Confidence intervals, on the other hand, estimate the parameter which is the proportion of individuals in the population or the probability that a randomly selected individual will fall within a certain range.

Comparisons of the observed frequencies of the three phenotypes of band 2 and the frequencies expected on the basis of Hardy-Weinberg analysis indicated that many of the hatchery populations were not in equilibrium while all of the natural populations which did not have the LP_2 allele fixed were in Hardy-Weinberg equilibrium. One possibility for the deviation from expected values could be misclassification of the quantitative phenotypes, but this can be considered to be negligible since good fits to expected values were observed in the genetic analysis of progenies of parents of predetermined lens protein phenotypes (Eckroat and Wright, 1969). In order for the Hardy-Weinberg Law to be valid one must assume random breeding, no mutation, no selection, no migration, and a large population size. It might be unreasonable to assume that all of these conditions are met in any natural or hatchery population. In the analysis of hatchery brook trout populations the effects of migration, mutation and population size can be considered to be minimal. Thus, selection or nonrandom breeding may be causing the deficiency of heterozygotes (Table 1) that is causing the deviation from Hardy-Weinberg equilibrium. It cannot be inferred that this implies selection for the homozygotes or against the heterozygotes; this analysis gives no values as to the relative fitness of the genotypes.

Artificial selection, that is selection resulting from the action of the breeder on the choice of parents, is probably operative in hatchery populations of brook trout. This type of selection may be based upon such factors as time of spawning, size of fish, and longevity. A degree of inbreeding within the population would probably result from this type of selection. Wright's Equilibrium Law which is a generalization of the Hardy-Weinberg Law (for which inbreeding $= 0$) takes into account the effects of inbreeding on zygotic proportions (Wright, 1922). Since inbreeding is completely independent of gene frequencies, the gene frequencies indicate the proportion of each lens protein allele in a population, while inbreeding influences how they associate in pairs. Thus, artificial selection can lead to inbreeding which would cause an increase in homozygotes at the expense of the heterozygotes while the gene frequency remains unchanged.

If artificial selection and inbreeding are causing deviations from Hardy-Weinberg equilibrium, one might expect the LDH and serum transferrin alleles to be affected similarly to the lens protein alleles unless selection is directed toward the lens protein alleles. The LDH and serum transferrin systems of these same hatchery populations satisfy Hardy-Weinberg Equilibrium (Wright and Atherton, 1970). Thus, no definite statement as to the failure of the lens proteins to satisfy Hardy-Weinberg equilibrium can be made; there are many factors involved which cannot be dealt with by means of a gene frequency analysis.

The lens protein polymorphisms in the natural populations of brook trout were similar to those in hatchery fish except that one does not find the range of allele frequencies observed in the latter. In all natural populations the LP_2 allele was either at or near fixation, while the LP_4 allele ranged from 0 to 0.23. All populations except that in Long Hollow had the LP_5 allele fixed. Inter-populational heterogeneity tests indicated that the dominant allele frequencies of the lens protein loci can be used to distinguish some, but not all, of the natural populations sampled in this study. The Long Hollow population was the most diverse of the natural populations. The location and nature of this stream in relation to the other streams sampled in this study makes it the most likely to be stocked with hatchery fish by local sportsmen.

It was observed that wild brook trout had much higher concentrations of proteins in a lens of equal size than did hatchery brook trout. This is presumably due to the fact that a wild brook trout of a certain age is smaller and has a smaller lens than a hatchery brook trout of the same age. This agrees with studies in mammals which demonstrated that lens growth was greater in pen-reared than in wild individuals (Lord, 1962). This higher concentration of lens proteins may account for the appearance of bands 6a, 8a and 8b in natural but not in hatchery populations. These bands do not appear in all wild fish. This finding of variants in wild populations is similar with the finding of LDH mutants in wild populations, a point stressed by Wright and Atherton (1970).

Consideration of the barrier imposed upon the Columbia County populations by the acid mine drainage into Catawissa Creek suggests that what at one time may have been a single base population is now subdivided into subpopulations. A subdivision into small populations should result in a dispersive process (Falconer, 1960). The dispersive process, if operative, leads to irregular fluctuation in gene frequency as the populations spread apart progressively and thus become differentiated. The limits imposed on this dispersion lead to fixation or loss of a particular allele. The fact that there is no differentiation of the Columbia County brook trout populations tends to rule out any effects of dispersion due to small population size unless there is also a strong heterozygote superiority counterbalancing the effects of dispersion. Even so, one would not expect all of the populations to be fixing the same alleles. However, no definite statement as to the effects of dispersion can be made since the allele frequencies of the base population are not known.

Since there is little or no detrimental effect on fish growth or survival after one eye lens is removed, at least in a hatchery environment (Eckroat and Wright, 1969), the lens proteins can be used in genetic analyses of populations without destroying the population. This is an obvious advantage over tissue systems where population samples must be killed. All of the lens protein bands which have been genetically characterized are present in a fish 125 days after hatching (1–2 inch fish); thus, within reason, no ontogenetic effects will hinder a lens protein analysis.

ACKNOWLEDGMENTS

The writer wishes to thank the following for their contributions to this study: Dr. James E. Wright, Jr. for directing the study and aiding in preparation of the manuscript; Louisa M. Atherton and the biologists and technical staff at the Benner Spring Fish Research Station for technical assistance. Specimens for this study were made available by Mr. Harvey Beall and Mr. David Cochran of the West Virginia Game and Fish Division and Mr. C. B. Corson and Mr. Charles Evans of the New Hampshire Department of Fish and Game.

LITERATURE CITED

BARRETT, I., AND H. TSUYUKI. 1967. Serum transferrin polymorphism in some scombroid fishes. Copeia 1967(3): 551–557.

———, AND A. WILLIAMS. 1967. Soluble lens proteins of some scombroid fishes. Copeia 1967(2): 468–471.

DAVIS, B. J. 1964. Disc electrophoresis II. Method and application to human serum proteins. Ann. New York Acad. Sci. 121: 404–427.

ECKROAT, L. R., AND J. E. WRIGHT, JR. 1969. Genetic analyses of soluble lens protein polymorphism in brook trout (Salvelinus fontinalis). Copeia 1969 (3): 466–473.

FALCONER, D. S. 1960. Introduction to Quantitative Genetics. Ronald Press, New York, pp. 47–50.

FUJINO, K. 1970. Immunological and biochemical genetics of tunas. Trans. Amer. Fish. Soc. 99(1): 152–178.

———, AND T. KANG. 1968a. Transferrin groups of tunas. Genetics 59: 79–91.

———, AND ———. 1968b. Serum esterase groups of Pacific and Atlantic tunas. Copeia 1968(1): 56–63.

———, AND T. K. KAZAMA. 1968. The y system of skipjack tuna blood groups. Vox Sang 14: 383–395.

KOEHN, R., AND D. RASMUSSEN. 1967. Polymorphic and monomorphic serum esterase heterogeneity in catostomid fish populations. Biochem. Genetics 1(2): 131–144.

LEITRIZ, E. 1959. Trout and salmon culture (Hatchery methods). State of California, Dept. of Fish and Game. Fish Bull. 107: 134–137.

Lord, R. D., Jr. 1962. Aging deer and determination of their nutritional status by the lens technique. Proc. 1st Nat'l White Tailed Deer Symposium, pp. 89–93.

Marr, J. C. 1957. The problems of defining and recognizing subpopulations of fishes. *In* J. C. Marr (coord.), contributions to the study of subpopulations of fishes, pp. 1–6. U. S. Fish Wildl. Serv., Spec. Sci. Rept. Fish. 208, 129 pp.

———, AND L. M. Sprague. 1963. The use of blood group characteristics in studying subpopulations of fishes. Int. Comm. Northwest. Atl. Fish., Spec. Publ. 4: 308–313.

Ornstein, L. 1964. Disc electrophoresis I. Background and theory. Ann. New York Acad. Sci. 121: 321–349.

Pearson, E. S., AND H. O. Hartley. 1954. Biometrika tables for Statisticians. Vol. I. Cambridge University Press.

Smith, A. C., AND R. A. Goldstein. 1967. Variation in protein composition of the eye lens nucleus in ocean whitefish, *Caulolatilus princeps.* Comp. Biochem. Physiol. 23: 533–539.

Wright, J. E., Jr., AND L. M. Atherton. 1970. Polymorphism for LDH and transferrin loci in brook trout populations. Trans. Amer. Fish. Soc. 99(1) : 179–192.

Wright, S. 1922. Coefficients of inbreeding and relationship. Am. Naturalist 56: 330–338.

Wright, T. D., AND A. D. Hasler. 1967. An electrophoretic analysis of the effects of isolation and homing behavior upon the serum proteins of the white bass (*Roccus chrysops*) in Wisconsin. Am. Naturalist 101: 401–413.

AUTHOR INDEX

Atherton, Louisa M., 9, 105
Avise, John C., 129

Champion, Michael, 110
Childers, William F., 110

Davisson, Muriel Trask, 105

Eckroat, Larry R., 91, 149

Hershberger, William K., 79

Kitto, G. Barrie, 129

Lewis, Robert D., 39

Morrison, William J., 49

Ridgway, George J., 39

Sherburne, Stuart W., 39

Tranquilli, John, 110

Whitt, Gregory S., 110
Wright, James E. Jr., 9, 105

KEY-WORD TITLE INDEX